Springer Theses

Recognizing Outstanding Ph.D. Research

Aims and Scope

The series "Springer Theses" brings together a selection of the very best Ph.D. theses from around the world and across the physical sciences. Nominated and endorsed by two recognized specialists, each published volume has been selected for its scientific excellence and the high impact of its contents for the pertinent field of research. For greater accessibility to non-specialists, the published versions include an extended introduction, as well as a foreword by the student's supervisor explaining the special relevance of the work for the field. As a whole, the series will provide a valuable resource both for newcomers to the research fields described, and for other scientists seeking detailed background information on special questions. Finally, it provides an accredited documentation of the valuable contributions made by today's younger generation of scientists.

Theses are accepted into the series by invited nomination only and must fulfill all of the following criteria

- They must be written in good English.
- The topic should fall within the confines of Chemistry, Physics, Earth Sciences, Engineering and related interdisciplinary fields such as Materials, Nanoscience, Chemical Engineering, Complex Systems and Biophysics.
- The work reported in the thesis must represent a significant scientific advance.
- If the thesis includes previously published material, permission to reproduce this must be gained from the respective copyright holder.
- They must have been examined and passed during the 12 months prior to nomination.
- Each thesis should include a foreword by the supervisor outlining the significance of its content.
- The theses should have a clearly defined structure including an introduction accessible to scientists not expert in that particular field.

More information about this series at http://www.springer.com/series/8790

Ang Yan Shan

Engineering a Robust DNA Circuit for the Direct Detection of Biomolecular Interactions

Doctoral Thesis accepted by
National University of Singapore, Singapore

Springer

Author
Dr. Ang Yan Shan
Department of Chemical
and Biomolecular Engineering
National University of Singapore
Singapore, Singapore

Supervisor
Assoc. Prof. Lanry Lin-Yue Yung
Department of Chemical
and Biomolecular Engineering
National University of Singapore
Singapore, Singapore

ISSN 2190-5053 ISSN 2190-5061 (electronic)
Springer Theses
ISBN 978-981-13-2187-0 ISBN 978-981-13-2188-7 (eBook)
https://doi.org/10.1007/978-981-13-2188-7

Library of Congress Control Number: 2018953186

This Springer imprint is published by the registered company Springer Nature Singapore Pte Ltd.
The registered company address is: 152 Beach Road, #21-01/04 Gateway East, Singapore 189721, Singapore

Supervisor's Foreword

The area of dynamic DNA nanotechnology, or DNA circuit, utilizes the well-established Watson–Crick base pairing rules to engineer different types of DNA molecular machines. It holds great promises as a modular-based and highly programmable toolbox for diverse applications including molecular computing and biomolecular detection. One bottleneck hindering the experimental implementation of in silico design is the recurring issue of circuit leakage due to spurious hybridization events which contribute to background noise.

Dr. Ang Yan Shan's thesis addresses the issue by designing a localized DNA circuit to promote the rate of desired reaction(s) over undesired leak reactions in the bulk solution. The formation of the localized circuit is driven by the recognition of biomolecules which is a dynamic assembly process. This differs from other approaches using static DNA origami assembly which enhances the hybridization kinetics of both the desired and leak reactions. There are two main themes in this thesis. First, individual module of the circuit, i.e., association toehold for turn-on signal transduction and hybridization chain reaction (HCR) for isothermal signal amplification was studied and modified independently. Next, the modules were integrated as a "plug-and-play" circuit for biosensing applications.

There are three key scientific contributions from this thesis. First, a general set of guidelines for designing robust HCR hairpins was established based on NUPACK simulation and experimental validation. Next, the association toehold design was modified by a simple trick to translocate a single nucleotide across the split domain, termed "inter-domain bridging", to achieve triple positive outcomes of enhanced hybridization rate across the three-way junction, reduced circuit leakage and overall improvement in the signal-to-noise ratio. Finally, direct experimental evidence was provided to show that oligonucleotide truncated at the 3′-end is the key source of synthesis defect. This contrasts with the commonly held notion in the field that truncation at the 5′-end is the main source of defect and offers an explanation for previously reported observations. The findings in these design concepts will be useful for other researchers, especially those unfamiliar with this field, to adopt DNA circuit as a toolbox without the need for empirical trial-and-error.

Eventually, the design concepts were successfully applied for biomolecular sensing of a broad range of targets including nucleic acids mutation, protein, and cell surface receptors. This demonstrates the generality of the "plug-and-play" platform, which we termed as the split proximity circuit and paves the way for developing the platform for practical detection applications. A proof-of-concept example for detecting protein complexes associated with breast cancer is provided in Chap. 8. Overall, this thesis is a comprehensive study which takes a DNA circuit design from concept development to applications on biological samples.

Singapore
August 2018

Assoc. Prof. Lanry Lin-Yue Yung

Abstract

DNA is the basic building block of life with well-characterized thermodynamics of hybridization. The predictability of Watson–Crick base pairing rules has inspired the use of DNA as a material to design different molecular structures and devices. The field of dynamic DNA nanotechnology, or DNA circuit, focuses on designing kinetic pathways, which are mostly driven by toehold-mediated strand displacement reactions, for autonomous computation. Despite its promises as a highly programmable toolbox, DNA circuit has yet to mature into a widely adopted method in applications such as bio-detection and imaging.

A key bottleneck is the recurring issue of circuit leakage, along with the lack of explicit design guidelines for the practical implementation of each design concept. This thesis aims to address this issue by designing a localized DNA circuit to promote the rate of desired reaction(s) over undesired leak reactions in the bulk solution. The recognition of biomolecules and their interaction events was used to trigger the dynamic assembly of the localized circuit.

We first established the design framework of a self-contained circuit consisting of three modules, i.e. target recognition, signal transduction, and signal amplification, using α-thrombin as a model biomolecule. The effect of localization on the rate of desired reaction was confirmed by varying the spacer length of the initiator strands. Individual modules of association toehold for signal transduction and hybridization chain reaction (HCR) for signal amplification were next studied in detail with a focus on establishing design concepts and guidelines to minimize their respective leakages.

A set of four-point design guideline for obtaining robust HCR hairpin sequences was identified via NUPACK simulation and validated experimentally. A new pair of hairpins with shorter stem length was generated with faster hairpin opening kinetics and effective Förster resonance energy transfer (FRET) readout signal. The HCR FRET system exhibited rapid kinetics, tunable detection range, and performed robustly in a biological environment. Next, the sources of leakages in association toehold, where all critical domains were exposed in active, single-stranded form, were identified. Two counter-leakage concepts were established—first, 3′-truncated synthetic oligonucleotides was identified as the main culprit of initial leakage burst and next, nucleotides were translocated across domains, in a strategy termed "inter-domain bridging", to suppress leakage and boost the strand displacement kinetics across a three-way junction.

Finally, the optimized modules were integrated into the final design, named "split proximity circuit". We demonstrated that the circuit could be dynamically assembled by various biomolecules and their interaction events in a "plug-and-play" format. The circuit was successfully applied for detecting nucleotide mutations in DNA and microRNAs, estimating binding affinity between α-thrombin and its recognition aptamer, as well as visualizing clustering events between cell receptors. It operated autonomously in a one-pot format, with good selectivity, sub-nanomolar limit of detection, and reasonable time scale of approximately 30 min. Deliberate efforts to minimize circuit leakages played a key enabling role in the successful implementation of this circuit design, which is now well-positioned to tackle real applications, such as for the clinical evaluation of breast cancer biomarkers.

Parts of this thesis have been published in the following journal articles:

1. Ang, Y. S., Li, J. J., Chua, P. J., Ng, C. T., Bay, B. H. and Yung, L.-Y. L. (2018) Localized Visualization and Autonomous Classification of Cell Surface Receptor Clusters Using DNA Proximity Circuit. *Anal. Chem.*, **90**, 6193–6198.
2. Ang, Y.S., Tong, R. and Yung, L.-Y.L. (2016) Engineering a Robust DNA Split Proximity Circuit With Minimized Circuit Leakage. *Nucleic Acids Res.*, **44**, e121.
3. Ang, Y.S. and Yung, L.-Y.L. (2016) Rational Design of Hybridization Chain Reaction Monomers For Robust Signal Amplification. *Chem. Commun.*, **52**, 4219–4222.
4. Ang, Y.S. and Yung, L.-Y.L. (2014) Engineering Self-Contained DNA Circuit For Proximity Recognition and Localized Signal Amplification of Target Biomolecules. *Nucleic Acids Res.*, **42**, 9523–9530.

Declaration

I hereby declare that this thesis is my original work and it has been written by me in its entirety. I have duly acknowledged all the sources of information which have been used in the thesis.

This thesis has also not been submitted for any degree in any university previously.

December 2016 Ang Yan Shan

Acknowledgements

My deepest gratitude to my supervisor Prof. Lanry Yung for being the key guiding figure of my Ph.D. study. Thanks for all your patience during the brainstorming sessions, and for giving me the freedom to explore while asking thought-provoking questions for me to reflect on and keep on track. I hope I did not disappoint as I grew to steer the wheels of this DNA circuit project and will continue to grow and learn for other scientific endeavors, be it ongoing projects or in the future.

Thanks to my thesis panel members, Prof. Yang Kun-Lin and Prof. Xie Jianping, for asking critical questions and providing different perspectives to this project over the past few advisory meetings. I truly value all the comments which helped to shape the thesis in its current form.

Thanks to all my lab mates, both graduated seniors and current members including Dr. You Fang, Dr. Suhanya Duraiswamy, Dr. Ma Ying, and Dr. Li Weijia, for all the support, encouragement and little moments of fun and laughter over the years. Special mention to Dr. Jasmine Li Jia'En who, despite the short stint in our lab, played an important role in getting the cell works up and running.

Here is a shout-out to my students over the years: Ms. Zhang Min and Ms. Chua Hui Min for working hard on the cell-SELEX protocol though circumstances did not allow us to pursue this direction further, Ms. Rachel Tong and Mr. Daniel Wee Bingqiang for pushing on with the association toehold and HCR projects during "turbulent" times, Mr. Benjamin Spanos (UCL) for his useful HCR simulation work, Mr. Alex Sim Chen Siong, and Ms. Elysia Koh Hui Min for bringing much joy and laughter while working hard for the DNA circuit projects, and Mr. Elvin Woon Wei Wen for the synergy in churning out "crazy ideas" for the silver nanocluster work.

Thanks to National University of Singapore for having me for almost a decade and the financial support. Thanks to the ChBE admin office for always being so helpful and efficient. Thanks to the laboratory staff at the WS2 lab including Mr. Lim You Kang, Mr. Ang Wee Siong and Mr. Tan Evan Stephen, and Dr. Yang Liming and Ms. Li Xiang, for all the assistance rendered over the years. Thanks also to all other technical officers who helped in instrument operations and other discussions.

Thanks to my overseas collaborators who kindly hosted and guided me on new techniques: Prof. Nguyen T. K. Thanh, Ms. Roxanne Hachani, and other lab mates from University College London; Prof. Hiroshi Sugiyama, Prof. Toshikazu Bando, Mr. Sefan Asamitsu, Ms. Kaori Hashiya, and other lab mates from Kyoto University. Thanks also to Prof. Bay Boon Huat, Dr. Ng Cheng Teng, and Dr. Chua Pei Jou from NUS Department of Anatomy for all the assistance in the cell works.

Thanks to my friends for the chill outings or simply for staying in touch. The laughter we shared during the catch-up sessions were indeed therapeutic.

Lastly, I would like to thank my parents for their silent support over the years. Their simple gesture of ensuring that a warm dinner awaits me when I got off late from school meant much.

Contents

Abbreviations

A	Adenine
BINDA	Binding-induced DNA assembly
BM	Branch migration domain
bp	Base pair
BSA	Bovine serum albumin
C	Cytosine
CHA	Catalytic hairpin assembly
DNA	Deoxyribonucleic acid
ELISA	Enzyme-linked immunosorbent assay
ErbB	Human epidermal growth factor receptor
FBS	Fetal bovine serum
FL.	Fluorescence
F-Q	Fluorophore–quencher
FRET	Förster Resonance Energy Transfer
G	Guanine
GFP	Green fluorescent protein
HCR	Hybridization chain reaction
HER1–4	Human epidermal growth factor receptor 1–4
HER2	Human epidermal growth factor receptor 2
HER3	Human epidermal growth factor receptor 3
HP	Mixture of HP1 and HP2
HP1	Hairpin 1
HP2	Hairpin 2
I	Mixture of I1 and I2
I1	Initiator 1
I2	Initiator 2
LOD	Limit of detection
miRNA	MicroRNAs
mRNA	Messenger RNAs
MW	Molecular weight

NC	Negative control
nt	Nucleotide
PCR	Polymerase chain reaction
PLA	Proximity ligation assay
qRT-PCR	Quantitative real-time polymerase chain reaction
RCA	Rolling circle amplification
RFU	Relative fluorescence unit
RNA	Ribonucleic acid
rSI	Relative signal intensity
S.D.	Standard deviation
S/B	Signal-to-background ratio
SATP	Split aptamer-based activatable theranostic probe
ssDNA	Single-stranded DNA
Stav	Streptavidin
STED	Stimulated emission depletion
STORM	Stochastic optical reconstruction microscopy
T	Thymine
TH	Toehold domain
x	Branch migration length

Symbols

D	Diffusion coefficient (cm^2/s)
k'	Rate constant of pseudo first-order reaction (s^{-1} or min^{-1})
k	Hybridization rate ($M^{-1}\ s^{-1}$)
k_{ini}	Initial reaction rate
K_D	Dissociation constant (M)
n	Number of independent samples
T	Temperature (K)
T_M	Melting temperature (°C)
ΔG	Change in Gibbs free energy ($kJ\ mol^{-1}$)
ξ	Extent of reaction
λ	Wavelength (nm)
λ_{Ex}	Excitation wavelength (nm)
λ_{Em}	Emission wavelength (nm)
Φ_F	Fluorescence quantum yield
$\tau_{reaction}$	Characteristic time of reaction, or reaction half-life (s/min/h)

List of Figures

List of Tables

List of Schemes

Chapter 1
Introduction

1.1 Motivation

DNA circuit is a highly programmable toolbox based on the dynamic interactions between logic gates driven by a series of hybridization events known as toehold-mediated strand displacements. Since the circuit design is modular and based primarily on the well-studied Watson-Crick base pairing, it is theoretically possible to engineer both simple and complex circuit operations to operate autonomously with high degree of control. Yet, while much progress has been made in the in silico design and experimental implementation of specific design concepts, the pace at which they are translated into real applications is lagging.

The initial goal of DNA circuit was to develop DNA molecular machines to execute autonomous molecular computation or large-scale data storage. The field had since acknowledged that DNA computation can never replace our everyday electronic computers, primarily due to the much slower kinetics of DNA hybridization as compared to electron fluxes. Instead, the unique combination of biology and computer sciences positions DNA circuit as an excellent candidate for interfacing directly with biomolecular systems, either for detection and imaging (more common), or inducing biological effects (upcoming direction). Though many promising strategies have been successfully demonstrated on proof-of-concept models in individual laboratories, only a handful made it past benchtop experimentation to actual clinical applications.

One major bottleneck hindering this progress is the issue of circuit leakage which can be simply understood as spurious hybridizations taking place outside of the designed operation. This can either be an inherent issue arising from imperfectly-synthesized oligonucleotide or due to cross talks between different DNA strands, particularly when the helical ends "breathe" to momentarily expose the nucleotides in single stranded form. These effects are not adequately accounted for in simulation tools and any ill-informed design attempt often leads to an ineffective trial-and-error

A. Yan Shan, *Engineering a Robust DNA Circuit for the Direct Detection of Biomolecular Interactions*, Springer Theses,
https://doi.org/10.1007/978-981-13-2188-7_1

process. This forms a barrier to the adoption of DNA circuit design concepts by other researchers.

There is therefore a need to establish explicit design principles or guidelines to minimize the occurrence of circuit leakages. A less obvious strategy is to design circuits which favour the reaction rate of desired processes over undesired leak processes. This can be achieved by constructing a localized circuit which assembles dynamically in response to specific triggering event(s), e.g. recognition of biomolecules or biomolecular interactions, to favour the strand displacement kinetics.

The successful development of this design concept can help to overcome upstream design challenges including leakage and slow computation speed. From an application angle, the as-designed circuit can be used for the direct detection of biomolecules and their interaction events. Preferably, the design should be sufficiently general such that it can be used across different target types in a "plug-and-play" format that is simple for end users to execute.

1.2 Aim and Objectives

The overall aim of this thesis is to design a robust, localized DNA circuit which is dynamically assembled by biomolecules interacting in close proximity. This was achieved via three key objectives:

Objective 1: To design a target-responsive, self-contained DNA circuit capable of executing localized strand displacement reaction.

A localized circuit promotes the strand displacement kinetics (Sect. 2.3.5). Moreover, having all circuit components self-contained on the target(s) facilitates the direct visualization of the target recognition. This objective guided the design process which is elaborated in Chaps. 4–6.

Objective 2: To minimize circuit leakages through thermodynamically-driven design concepts.

Circuit leakage is a major problem in the field of DNA circuit (Sect. 2.5.2). As such, a recurring theme when refining the circuit modules in Chaps. 5 and 6 was to develop strategies to counter this issue.

Objective 3: To demonstrate the generality of our platform for the simple, one-pot detection of bi-recognition events.

Once the upstream circuit design was finalized, the next focus was to demonstrate the practical application of the split proximity circuit as a "plug-and-play" platform (Chaps. 7 and 8).

Approach: This thesis is driven primarily by the experimental execution of key design concepts, while NUPACK simulation was used as a secondary tool to guide the sequence design and provide additional insights to explain certain experimental observations.

1.3 Organization of Thesis

This thesis is organized into 9 chapters. After this introductory chapter, key concepts in DNA circuitry and its relevant applications are reviewed in Chap. 2. All experimental details are found in Chap. 3. The main research results are presented in Chaps. 4–8 (Fig. 1.1). The overall design framework of my proposed circuit, i.e. the three modules of target recognition, signal transduction and signal amplification, is introduced in Chap. 4. Due to the modular nature of DNA circuit, each module could be designed and optimized independently in the subsequent chapters with a focus on minimizing circuit leakage. In Chap. 5, guidelines for designing the hairpins of hybridization chain reaction (HCR) were established and a fluorescent-based readout was incorporated. In Chap. 6, a generic signal transduction strategy based on the concept of association toehold to convert the bi-molecular recognition events into a DNA trigger was developed. The individually designed modules were then integrated as the "split proximity circuit" which I formally introduce in Chap. 7.

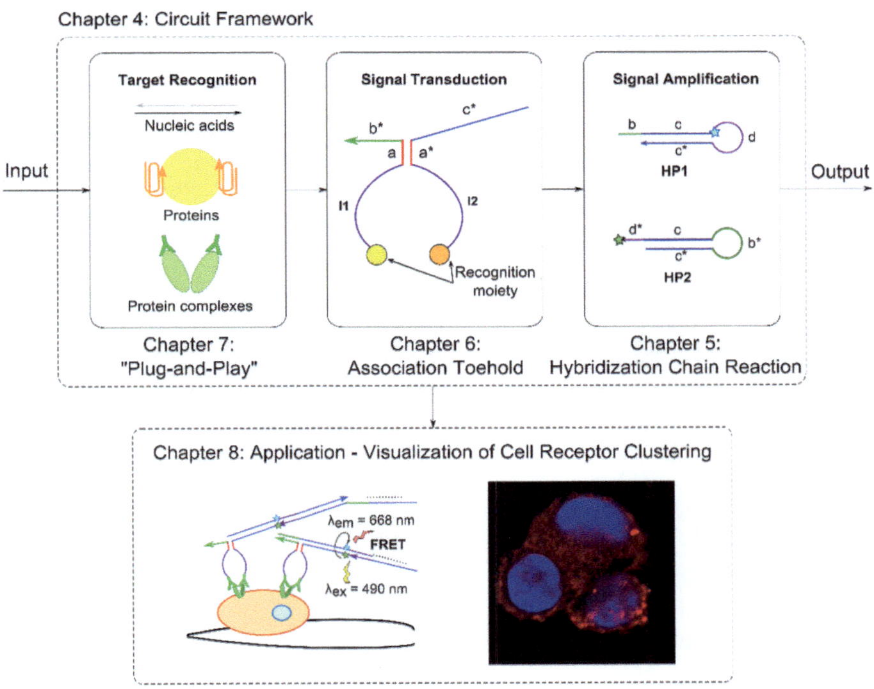

Fig. 1.1 Overview of main research results presented—from overview in Chap. 4 to individual module design in Chaps. 5–7, and finally a biological application in Chap. 8

In the same chapter, the "plug-and-play" format of the split proximity circuit was demonstrated using a range of biomolecular targets. A specific case study on using the split proximity circuit for visualizing cell surface receptor clustering was demonstrated in Chap. 8. This thesis concludes in Chap. 9 with suggestions of future works and a perspective on the future outlook of DNA circuit.

Chapter 2
Literature Review

The field of DNA circuitry started with the aim of developing DNA molecular machine for autonomous molecular computation. However, the field has since recognized that the speed and scale of DNA computation can never surpass that of silicon-based microelectronics [1]. Instead, the use of DNA molecule as a reactant renders it as an excellent candidate for interfacing with biomolecular systems which we will focus our review on [2]. We will first introduce key design concepts in DNA circuit from a modular perspective. The individual modules are then matched up in different combinations for applications in detecting and imaging of various -omics targets. Finally, we will briefly highlight some challenges for moving the field forward towards translational works.

2.1 Basic Introduction to DNA

2.1.1 DNA as the Building Block of Life

DNA is the blueprint of life encoded by basic repeating units called nucleotides. Each nucleotide consists of a phosphate group, five-carbon 2-deoxyribose sugar ring and a base (Fig. 2.1). There are four types of bases: adenine (A), thymine (T), cytosine (C) and guanine (G). The nucleotides are linked by covalent phosphodiester bond formed between the 3′-OH and 5′-phosphate group of adjacent nucleotides. The DNA backbone consists of alternating phosphate and sugar groups while the bases impart unique sequence identity to the DNA.

Two complementary DNAs hybridize with one another according to Watson-Crick base pairing rules to form a double helical structure [3]. There are two hydrogen bonds in a A·T base pair and three hydrogen bonds in a G·C base pair. While the difference in the number of hydrogen bond is often used to explain why G·C base pair is the stronger of the two, this simplified model often masks the more significant

A. Yan Shan, *Engineering a Robust DNA Circuit for the Direct Detection of Biomolecular Interactions*, Springer Theses,
https://doi.org/10.1007/978-981-13-2188-7_2

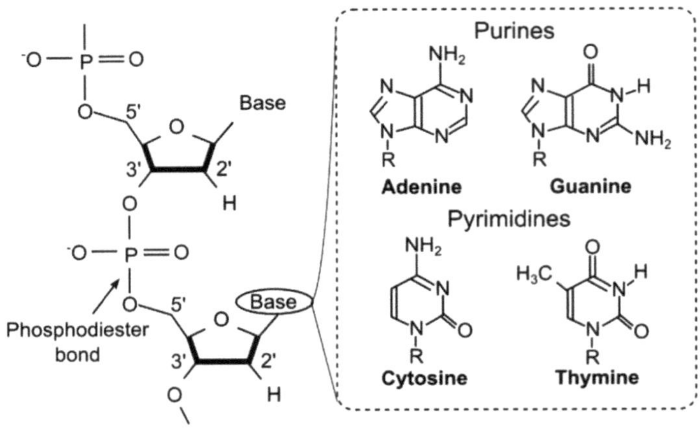

Fig. 2.1 The basic building block of DNA is nucleotide which consists of phosphate group, sugar ring and a base (in dotted box)

effect of base stacking between adjacent bases [4]. Essentially, the hydrogen bonds between the base pairs provides specific recognition and inter-strand stabilization for π-π interactions to occur between the aromatic rings of the bases which contribute further to both inter- and intra-strand stability [5, 6]. This coaxial stacking effect allows multiple shorter DNA strands to hybridize stably along a longer DNA strand.

It is worth mentioning that stacking also occurs in single stranded DNA (ssDNA), though with little cooperativity effect. Hairpin loop is a special case where complementarity in the short stem sequence, typically 4–6 nt, contributes to strong intramolecular hydrogen bonding for base stacking to take place. The equilibrium of hairpin formation is independent of DNA concentration while the hairpin free energy is the sum of parts from the stem and loop design [7]. The minimum loop size is three nucleotides, and an increase in the loop length destabilizes the hairpin [8–10].

2.1.2 *Kinetics and Equilibrium Models of DNA Hybridization*

The kinetics of DNA hybridization can be described with two elementary steps (Fig. 2.2). The first few bases bind to and initiate the nucleation of a duplex segment. This step is bimolecular and requires two complementary strand to diffuse, meet one another and associate successfully. The formation of the first base pair is a thermodynamically unfavourable process ($\Delta G > 0$) due to the lack of base stacking effect [11]. The rate constant is $10^7\ M^{-1}\ s^{-1}$ (for $[Na^+] = 1\ M$) for duplex nucleation and 10^4–$10^5\ s^{-1}$ for hairpin formation [12]. Once the first few (two to four) bases form successfully, the duplex structure is more stable and the next zippering step starts, where more base pairs are added to the existing helical region. This step is unimolecular with a rate constant of 10^6–$10^7\ s^{-1}$ and is thermodynamically favourable [13].

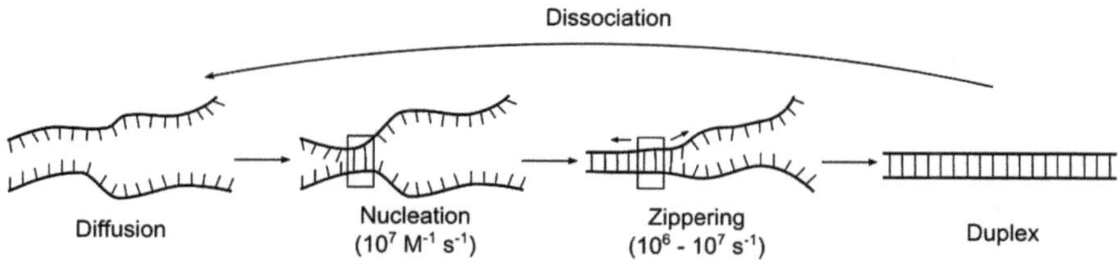

Fig. 2.2 Key steps involved in DNA hybridization. Two diffusing DNA strands bind at the first few bases to initiate duplex nucleation (slow step) followed by a fast zippering step to form a duplex structure. The dissociation rate is inversely proportionate to the DNA length

DNA hybridization is best represented by its melting temperature (T_M) which is defined as the temperature at which DNA is half denatured, i.e. 50% in duplex and 50% in single stranded form. It is modelled by the two-state melting theory for short oligonucleotides which assumes that the DNA is either in duplex or fully dissociated state [14]. The thermodynamics parameters, e.g. enthalpy, entropy [15] and Gibb's free energy, are predicted using the nearest neighbour model which accounts for both effects of base pairing and stacking interactions between adjacent nucleotides [14]. T_M depends on three factors: (1) sequence design—longer length and higher G·C content increases T_M, (2) oligonucleotide concentration—higher concentration favours duplex formation and increases T_M, and (3) salt environment—presence of Na^+ and Mg^{2+} influences DNA stability [16, 17].

2.1.3 Development of DNA Nanotechnology

DNA is known classically for its role in the central dogma for storing the genetic information of all living organisms [18]. Since the seminal work by Berg [19] in 1972 on the manipulation of genes across different species, the field of recombinant DNA has grown significantly and matured into what we know as biotechnology today. DNA nanotechnology takes it one step further by using synthetic DNA, instead of existing genetic sequences, as a starting material to design different molecular structures and devices.

The field started as structural DNA nanotechnology in 1982 when Seeman [20] first proposed that DNA junctions can be designed and assembled into networks of lattices through their sticky ends. Since then, several motifs have been designed and used as building blocks for assembly into two-dimensional and three-dimensional objects in various applications [21]. The second branch of the field, functional DNA nanotechnology [22], soon developed with the introduction of aptamers by Ellington and Szostak [23] in 1990. Aptamers are short oligonucleotides capable of binding to a wide range of targets with high affinity and specificity based on non-Watson-Crick interactions [24].

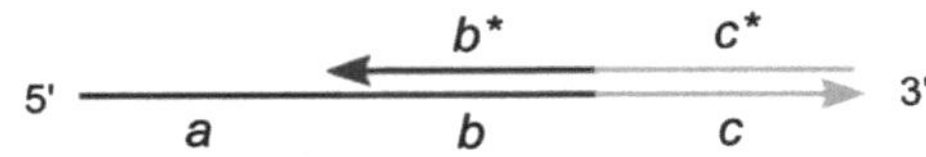

Fig. 2.3 Domain notation of DNA strands involved in circuit operation

Finally, the branch of dynamic DNA nanotechnology, or DNA circuit, emerged in 1994 when Adleman first introduced the use of molecular biology tools, including DNA strand displacement, for molecular computing to solve a travelling salesman mathematical problem [25]. The field took off in 2000 when Yurke and co-workers [26] introduced the concept of toehold, which is a short ssDNA overhang, to directly fuel the conformational changes in a simple three-stranded DNA tweezer structure. This design improved the strand displacement kinetics drastically and inspired many other enzyme-free molecular machine driven primarily by DNA hybridization [27].

2.2 Introduction to Dynamic DNA Nanotechnology (DNA Circuit)

Dynamic DNA nanotechnology focuses on programming the dynamics of DNA hybridization, rather than a single equilibrium end state, in response to specific input stimulus.

2.2.1 Nomenclature

The individual DNA strand is represented as a directional line drawn as an arrow from the 5′- to 3′-end (Fig. 2.3). Each DNA strand is divided into distinct segments, termed domain, which refers to a stretch of contiguous nucleotides forming a basic binding unit. The domains can be numbered or listed in alphabetical order or as Greek letters. Complementary domains are denoted by the same letter with an asterisk or overbar. We will bold the name of DNA strands, use italicized Roman alphabets to label the domains and denote complementary domains with an asterisk in this thesis.

2.2.2 Toehold-Mediated Strand Displacement

Almost all DNA circuits are powered by a simple concept of toehold-mediated strand displacement pioneered by Yurke and co-workers [26]. The concept is executed by two key domains—the toehold domain (domain a^*), which is a short DNA overhang from a duplex structure, and the branch migration domain (domain b^*), where strand displacement takes place (Fig. 2.4).

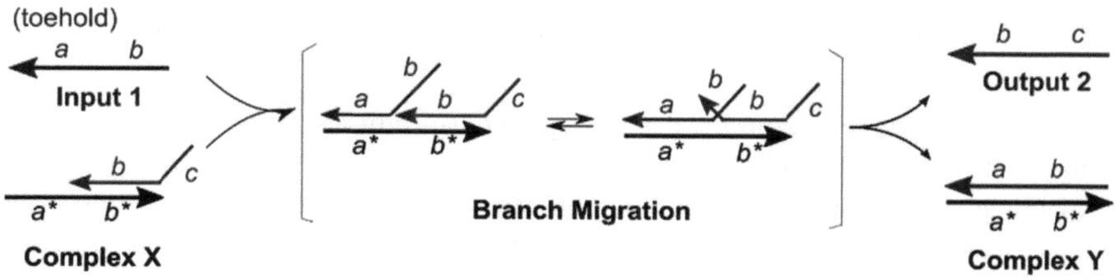

Fig. 2.4 Mechanism of toehold-mediated strand displacement. The incoming input strand binds to Complex X at the toehold, initiates branch migration which eventually displaces the output strand to form Complex Y

The invading strand (input 1) first binds to Complex X at the toehold (domain $a*$). Next, branch migration takes place across domain $b*$ in a series of nucleotide-by-nucleotide displacement at the branch point of the three-strand complexes. Eventually, input 1 invades Complex X successfully to release the incumbent strand (output 2) from the substrate and form a new Complex Y.

The thermodynamics of the overall reaction is driven by enthalpic and/or entropic gains. The former is more common where the gain in enthalpy is derived from the free energy released from toehold binding. The latter is more unique where a single input strand is used to release two or more output strands, resulting in a net gain in DNA species [28].

The toehold is the critical domain controlling both the thermodynamics and kinetics of strand displacement. In the absence of toehold, Högberg and co-workers [29] found that the strand displacement rate at 30°C was 3.6 and 12.8 M^{-1} s^{-1} for 12 and 16 nt probe respectively. Yurke and Mill [30] showed that an increase in toehold length from 0 to 6 nt increases the strand displacement rate by 6×10^4 M^{-1} s^{-1} while Zhang and Winfree [31] found that a maximum rate of 6×10^6 M^{-1} s^{-1} was achieved for toeholds longer than 7 nt. This is close to the hybridization nucleation rate presented in Sect. 2.1.2. The backward reaction is six orders of magnitude slower than the toehold-mediated reaction and the overall strand displacement reaction can be taken as irreversible.

Branch migration involves a series of base pairing and dissociation, both occurring with near equal probability, with each step forward (or backward). It is a random walk process which rate is dependent on the branch migration length (x) [32]. Zhang and Winfree [31] proposed that the branch migration rate can be approximated as $\frac{400}{x^2}$ s^{-1}. This empirical model was later refined by Winfree and co-workers [33] to better describe the process at the molecular level. The initiation of branch migration is a thermodynamically unfavourable process, where the first two base pairs contribute to the greatest energy penalty. Toward the end of branch migration, the incumbent strand can dissociate spontaneously from the substrate.

2.2.3 Simulation Tools to Facilitate Design of DNA Circuits

The final equilibrium state attained at the end of the circuit operation can be simulated using NUPACK [34], IDT SciTools [35] or Mfold [36]. All three thermodynamics packages are based fundamentally on the nearest-neighbour model described by SantaLucia (described in Sect. 2.1.2) with different degrees of modifications [37]. NUPACK concerns with the interactions between multi-strand nucleic acids; SciTools calculates key information, e.g. T_M and Gibb's free energy, based on the input sequence; while Mfold focuses on predicting the secondary structure of single stranded nucleic acids. NUPACK is the primary simulation tool for modelling DNA circuit operations since it considers the interaction dynamics between multiple DNA strands.

Other simulation tools include Visual DSD [38], which automates the computation of all possible reactions between large number of nucleic acids species, and OxDNA [39], which is a coarse-grained model for modelling processes involving DNA. These will not be used in this work which does not involve complex circuit design.

2.2.4 Advantages of DNA Circuit

DNA circuit is fundamentally based on DNA hybridization which is a well-established biophysical process. It offers a systematic framework for utilizing the known science to create a myriad of new processes and DNA molecular device designs. This unique positioning comes with several advantages.

First, DNA circuit is driven by Watson-Crick base pairing which is a well-studied phenomenon. The thermodynamics parameters are relatively well-established and so, the circuit can be designed in a highly predictable way. Moreover, the circuit design is modular and can be used in a "plug-and-play" format or to scale up the circuit complexity. The circuit can be programmed to operate autonomously using simple toehold-mediated strand displacement to communicate between different modules. The basic circuit design is enzyme-free and hence operates more robustly over a wide range of environmental conditions, e.g. temperature, pH and salt, and does not require co-factors to function.

2.3 Key Design Concepts in DNA Circuit

A single DNA circuit typically consists of several cascaded toehold-mediated strand displacement reactions to perform the required computation. We will look at different levels of design concepts with increasing complexity.

2.3.1 Toehold Designs

The classic way of achieving timed activation of strand displacement is by sequestering the toehold sequence with another complementary strand [40] or within secondary structures, e.g. hairpin loop [41]. The conventional strategies are driven by the unidirectional hybridization reaction between complementary DNA strands with little flexibility to fine-tune reaction kinetics due to the strong coupling between the kinetics and thermodynamics of strand displacement.

The two effects were decoupled using the concept of toehold exchange introduced by Zhang and Winfree [31] where the active and sequestered domains are exchanged upon strand displacement (Fig. 2.5a). This allows fast forward and reverse reactions to be designed even for sequences which are only mildly thermodynamically favourable. Another strategy is the remote toehold design introduced by Turberfield and coworkers [42] where a spacer is added between the toehold and branch migration domain (Fig. 2.5b). The overall strand displacement kinetics is now controlled by the additional rate-limiting step of internal diffusion and reduces into a first-order reaction with improved robustness against variations in the DNA concentration.

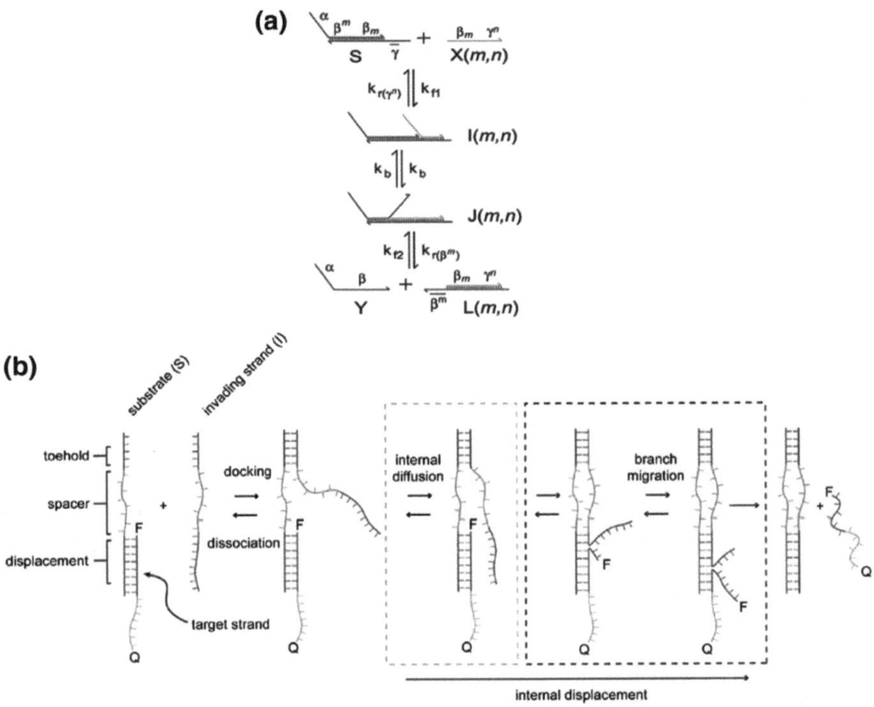

Fig. 2.5 DNA-based toehold activation using **a** toehold exchange (Adapted with permision from Ref. [31]. Copyright © 2009, American Chemical Society) and **b** remote toehold (Reprinted with permission from Ref. [42]. Copyright © 2011, American Chemical Society)

Besides using DNA strand as a trigger, alternative stimuli-responsive designs have emerged to respond to environmental changes or other types of biomolecule binding and offer finer control on the reaction kinetics. One popular strategy is to protect the toehold using aptamer sequences which reconfigures upon target binding to reveal the hidden toehold (Fig. 2.6a) [43–45]. However, the additional molecular reconfiguration step penalizes the kinetics appreciably where the overall reaction rates are reduced to 10^1–10^3 M^{-1} s^{-1}. Another approach utilizes the formation of known DNA secondary structures that responds to environmental stimulus, e.g. ions, pH and small molecules, to activate strand displacement instead of a toehold overhang

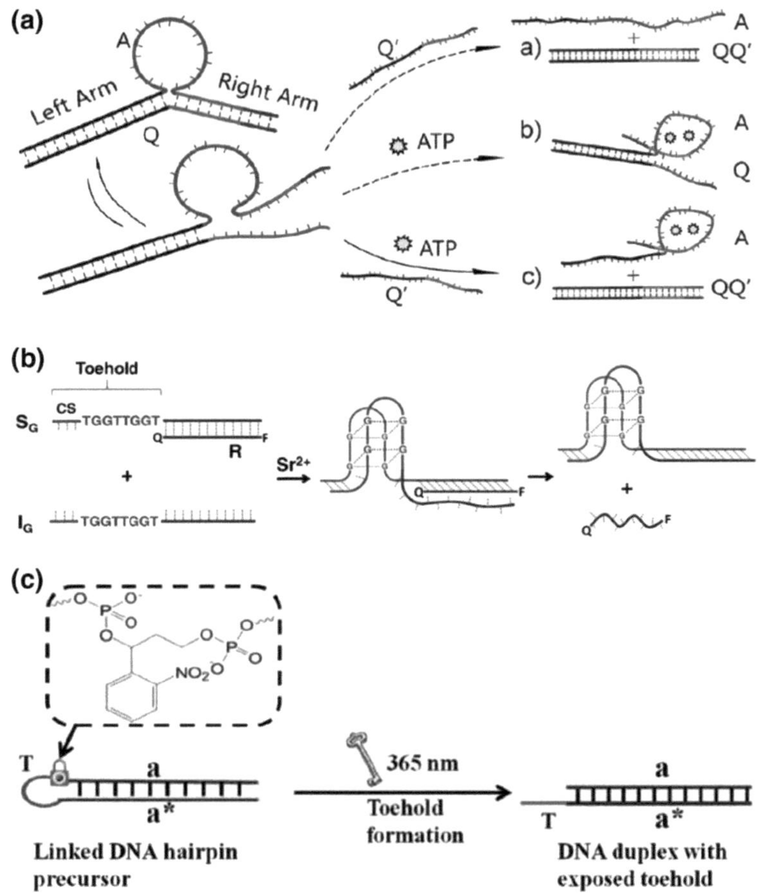

Fig. 2.6 Toehold activation based on **a** aptamer recognition (Reproduced with permission from Ref. [44]. Copyright © 2011, WILEY-VCH Verlag GmbH & Co. KGaA, Weinheim), **b** formation of DNA tetraplex (Reprinted with permission from Ref. [49]. Copyright © 2013, American Chemical Society) and **c** photocleavage upon light irradiation (Reprinted with permission from Ref. [52]. Copyright © 2013, American Chemical Society)

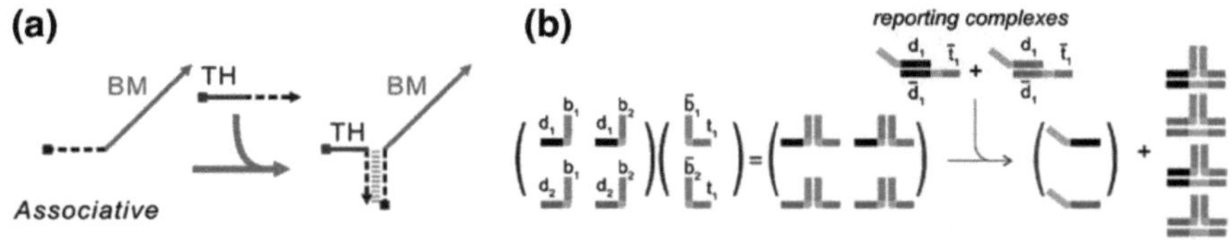

Fig. 2.7 **a** Association toehold mechanism (Reprinted with permission from Ref. [53]. Copyright © 2012, American Chemical Society), **b** where different combinations of toehold (TH) and branch migration (BM) domains can be used to trigger DNA combinatorial displacement (Reproduced with permission from Ref. [54]. Copyright © 2013, WILEY-VCH Verlag GmbH & Co. KGaA, Weinheim)

(Fig. 2.6b) [46–49]. The strand displacement rates can be tuned over two orders of magnitude by changing the stimulus concentration, akin to changing the toehold lengths. However, the absolute hybridization rates are noticeably slower than the conventional toehold activation, i.e. 10^2–10^4 M^{-1} s^{-1} for most of the designs.

A more general, faster and cleaner method is to activate toehold using light [50, 51]. Tan and co-workers [52] incorporated a photocleavable nitrobenzyl linker at the point of oligonucleotide synthesis at the loop of a hairpin. Upon irradiation with light ($\lambda = 365$ nm), the hairpin structure opens into a duplex structure and avails the toehold as a single stranded overhang (Fig. 2.6c). The strand displacement kinetics can be tuned by changing the irradiation time and 20 min of irradiation produces sufficient toehold for hybridization to proceed at a similar rate as the conventional strand displacement reaction. This method also can help to generate pure DNA duplexes from single stranded hairpins without further purification.

The activation strategies discussed thus far are based on dissociation mechanism or molecular reconfiguration. Xi [53] challenged the norm and demonstrated an alternative concept to the activation mechanism itself where the toehold and branch migration domain associates upon activation via DNA hybridization to trigger downstream circuit operation (Fig. 2.7a). This way, the two domains are not hardwired at the point of oligonucleotide synthesis and can be swapped around dynamically during circuit operation. The overall hybridization rate is 10^4–10^5 M^{-1} s^{-1} which is about an order of magnitude lower than the usual toehold-mediated strand displacement. Turberfield and co-workers [54]. takes this concept further to show how n toehold strands and n branch migration strands can be combined to generate n^2 complete strand displacement complexes, which they termed DNA combinatorial displacement (Fig. 2.7b). As such, a reduced set of DNA strands can be used to power more complex computations, such as to build a majority logic circuit which generates an output only if 50% of its inputs are true [55].

Overall, the conventional toehold design gives the fastest strand displacement kinetics (10^6 M^{-1} s^{-1}) but is restrictive from the design perspective since it is only triggered by DNA inputs and the toehold length controls both the kinetics and thermodynamics of hybridization. Alternative designs often penalize the reaction kinetics but value-add by improving circuit robustness, expanding the pool of target stimuli or providing finer control on the kinetics pathway.

2.3.2 Translator Module

An advantage of DNA circuit is its modularity where the domain sequences can, in theory, be designed independently of one another. This remains true only in a purely synthetic system with no sequence constraints. The design space for practical applications is likely to be restricted by the target sequences to be recognized. This implies that new sequences need to be generated for individual target which is inefficient. One solution is to incorporate a translator module to transduce any target(s) of interest, including non-nucleic acids targets, into a generic DNA output that is compatible with existing circuit sequence designs.

Ghadiri and co-workers [56] reported the design of a simple universal translator module for transducing input DNA sequences and other targets into the same output strand for signal generation. Two strand displacement steps are involved where the input strand binds to the first translator complex (first-half translation) and release the intermediate strand which then displaces the output on the second-half translator (Fig. 2.8a). The method is simple but effective where 74% yield of output was achieved within just 5 min of operation using a model hepatitis C virus sequence. We noticed the relatively high concentration of DNA components (10^{-7}–10^{-5} M) used in this proof-of-concept work. It remains to be seen whether the high yield obtained within this short incubation time can be reproduced under more dilute conditions. Other target types can also be recognized by using aptamers to sequester the first toehold.

Instead of adding more strand displacement step(s), a translator that reconfigures upon target binding to bring the toehold and branch migration domains together can be designed (Fig. 2.8b) [57–59]. The output DNA is used to trigger downstream readout modules, e.g. fluorophore-quencher and signal amplification readouts, or to perform logical computation. This strategy involving molecular reconfiguration is attractive but is restricted by the limited pool of targets known to interact directly with DNA to induce the conformation change needed in this strategy.

A more general translator design driven solely by the physical binding between the DNA circuit and target was introduced by Le and co-workers [60]. Gold nanoparticle was used as a scaffold to construct the DNA circuit gates (Fig. 2.8c). Upon target binding, the invading strand comes into proximity with the circuit gate to displace the output strand. Although the authors claimed that the gold nanoparticle helps to increase the local DNA concentration for improved equilibrium signal, it is not entirely clear how this effect has a bearing on the reaction between three separate entities (target, gate on nanoparticle and incumbent strand) moving freely in solution. Nonetheless, the concept was successfully applied for streptavidin-biotin and aptamer binding, which can be further adapted for protein targets via antibody-DNA conjugates [61].

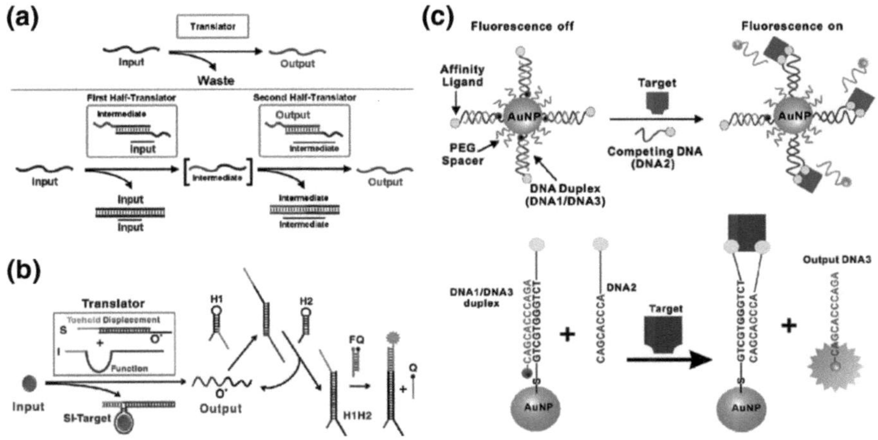

Fig. 2.8 Design of translator to convert different types of target into a generic DNA output *via* **a** two-step toehold-mediated strand displacement (Reprinted with permission from Ref. [56]. Copyright © 2009, American Chemical Society), **b** binding-induced molecular reconfiguration (Reproduced from Ref. [57] with permission from the Royal Society of Chemistry) and **c** binding-induced strand displacement (Reproduced with permission from Ref. [60]. Copyright © 2012, WILEY-VCH Verlag GmbH & Co. KGaA, Weinheim)

2.3.3 Signal Amplification Through Dynamic Self-assembly

DNA circuit can be engineered to perform enzyme-free signal amplification via dynamic self-assembly (enthalpy-driven) or entropy-driven pathways. The enzyme-free signal amplification concepts appear to be more readily adapted by other research communities particularly for biosensing and is a promising direction for further refinement and development.

Dirk and Pierce [62] first demonstrated the use of hairpin monomers to fuel enzyme-free signal amplification in a strategy termed hybridization chain reaction (HCR). Potential energy is stored within a pair of kinetically-locked hairpin loop which is released upon hybridization to a trigger strand (Fig. 2.9a). Each hairpin has two orthogonal toehold domains which binds to one another in a cascaded manner to polymerize into a long linear chain. The basic HCR concept can be expanded by increasing the number of hairpin monomers [63] or the number of toeholds per hairpin [64] to generate an extensive hyperbranched DNA structure (Fig. 2.9b).

Instead of trapping the trigger strand within the assembly structure, Pierce and co-workers [65] used hairpin monomers to programme the catalytic self-assembly of DNA over a range of designs, e.g. catalytic hairpin amplification (CHA), autocatalysis and dendritic growth of DNA assembly. In the simplest CHA design, the first hairpin opens upon binding to the catalytic stand which is released by the binding of the second hairpin to regenerate the catalytic strand (Fig. 2.10a). Multiple assembly reactions can be stacked to achieve exponential amplification from a single catalytic

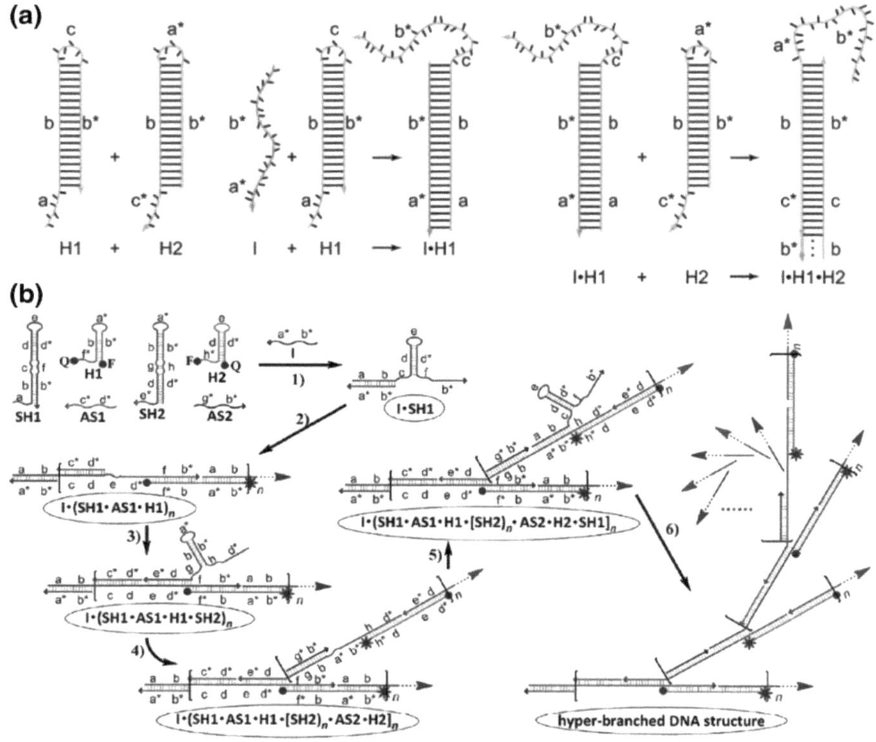

Fig. 2.9 Dynamic signal amplification using hairpin fuels based on **a** hybridization chain reaction (HCR) (Reproduced with permission from Ref. [62]. Copyright © 2004, National Academy of Sciences, U.S.A.) and **b** hyperbranched HCR (Reproduced with permission from Ref. [63]. Copyright © 2015, WILEY-VCH Verlag GmbH & Co. KGaA, Weinheim)

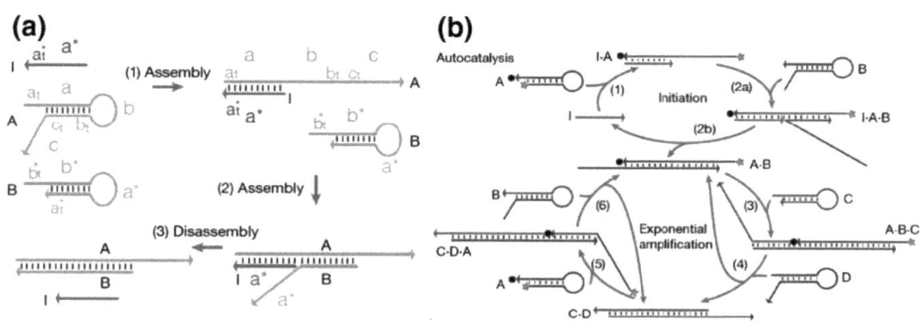

Fig. 2.10 Dynamic signal amplification based on **a** catalytic hairpin assembly (CHA) and **b** higher order CHA reaction leading to an autocatalysis reaction (Reproduced with permission from Ref. [65]. Copyright © 2008, Springer Nature)

strand (Fig. 2.10b). However, as the assemblies get increasingly complicated, the issues of circuit leakage and slow circuit kinetics become more significant.

The intermediate complexes involved in hairpin-fuelled amplification are thermodynamically less stable since the displaced domains are held near the new complex by the hairpin loop with higher probability of reversing the strand displacement process. Xuan and Hsing [66] instead used two duplexes, each with two sequestered toeholds, to drive the exponential growth of DNA dendrimers (Fig. 2.11). The rationale is that assemblies formed from duplex monomers are more stable than hairpin monomers which helps to drive the forward reaction. Significant signal can still be observed

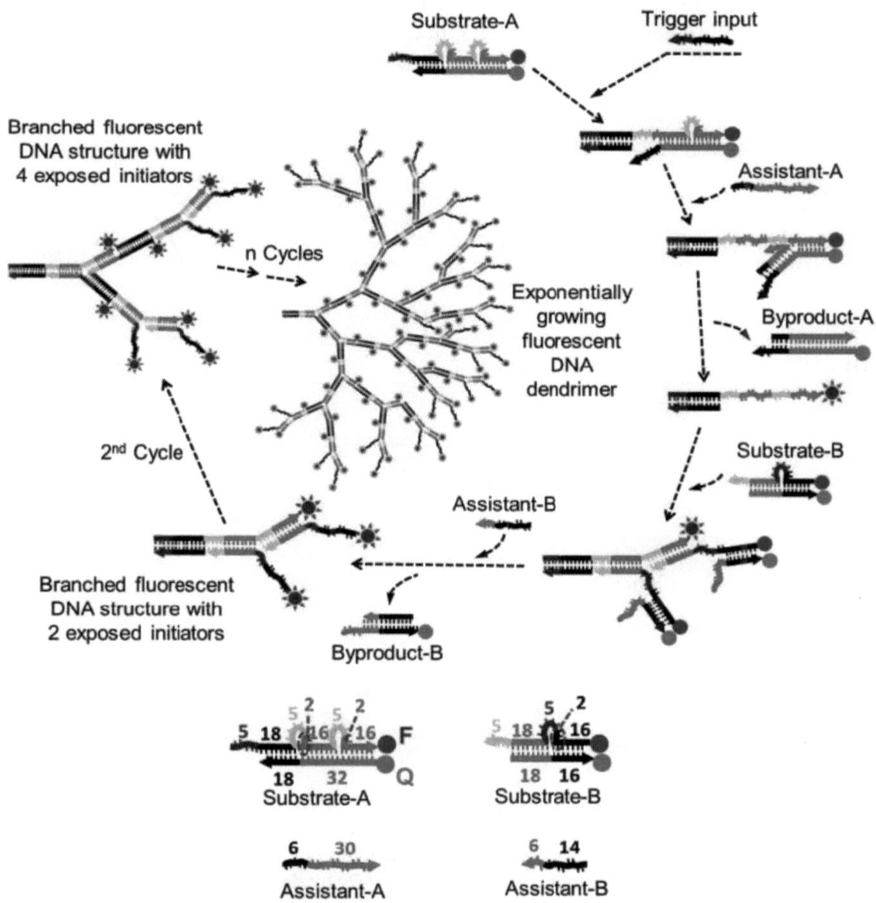

Fig. 2.11 A hairpin-free approach towards dynamic self-assembly of DNA dendrimers using DNA duplex fuels (Reprinted with permission from Ref. [66]. Copyright © 2014, American Chemical Society)

for low trigger concentration down to 0.001X DNA even though the presence of imperfectly formed duplexes does contribute appreciably to leak reactions.

2.3.4 Higher Order Gate Assemblies

The modularity of DNA circuit allows for scaling up of operations via a network of logic gates to perform complex computation. Winfree and co-workers [40] first designed different enzyme-free gates such as AND, OR, NOT and threshold, which can be used as building blocks to design elaborate strand displacement cascades (Fig. 2.12). They successfully implemented a five-layer circuit which takes in 6 inputs and uses 11 gates for computation. However, as the circuit increases in scale,

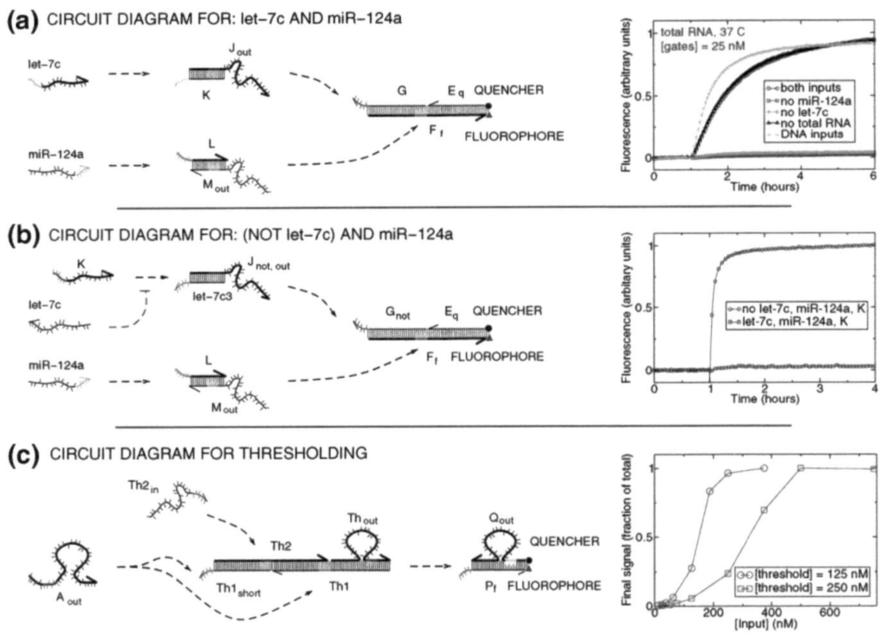

Fig. 2.12 Examples of DNA logic gate designs including **a** AND, **b** NOT and AND, and **c** threshold. The toehold sequestered in the gate complex are released upon target binding (complementary toehold domains are denoted by the same colour) to generate a output signal based on the combination of inputs recognized. For example, in **a** AND gate, let-7c binds to strand K *via* the blue toehold to release the brown toehold in strand J_{out}, which binds to and releases strand G from the readout complex. The sequestered pink toehold in F_l is revealed. The second target, miR-124a, binds to strand L to release the pink toehold in M_{out} which can then displace strand F_l from the quencher strand (E_q) *via* the newly available pink toehold. (From Ref. [40]. Reprinted with permission from AAAS)

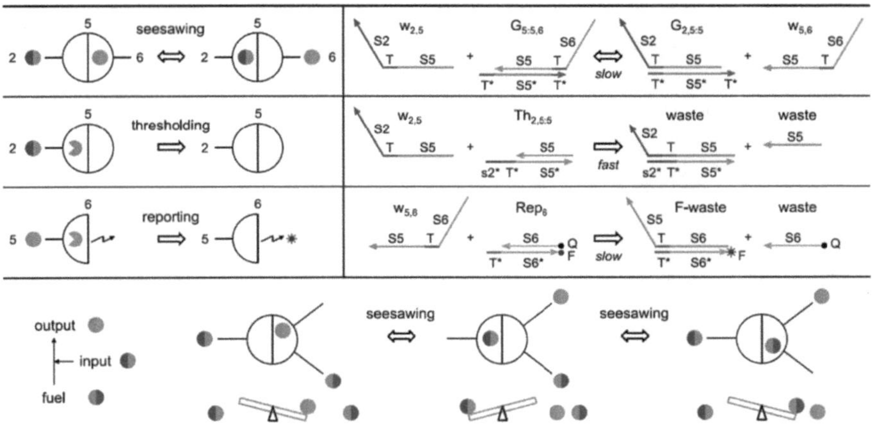

Fig. 2.13 A seesaw circuit consists of three key modules of seesaw (for transducing input strand), thresholding (to set the minimum input concentration required) and reporting (e.g. using fluorophore-quencher complexes) (From Ref. [68]. Reprinted with permission from AAAS)

computation becomes slower, e.g. the reaction half-life of the complete circuit was characterized to be 10 h.

An important concept of seesaw circuits was later introduced by Qian and Winfree [67] as a molecular compiler to convert high-level DNA programming language into the molecular implementation of DNA sequences. It consists of three basic modules of see-saw gate (based on the principle of toehold exchange), threshold and signal reporting (Fig. 2.13). A large-scale network is constructed via the diffusion of strands released from one see-saw gate to another to execute AND and OR logical computation. The most complex circuit utilizes 130 DNA strands to compute the square roots of 0, 1, 4 and 9 [68]. However, this design is a feedforward system for one-time computation and further scalability of the current solution phase operation is limited by the presence of spurious reactions and slow reaction rate (time scale of hours).

2.3.5 *Localized DNA Circuit*

A major barrier to the scale up of DNA circuit is its slow operation since DNA strands need to find one another in a dilute solution before the toehold can bind and initiate strand displacement [69]. One potential solution is to use localized DNA circuit design to speed up strand displacement kinetics and compartmentalize different parts of the circuit operation [70, 71].

To this end, Qian and Winfree [68] suggested using DNA origami to organize the circuit components such that interactions occur directly between adjacent gates rather than via diffusion. It also provides spatial compartmentalization for individual

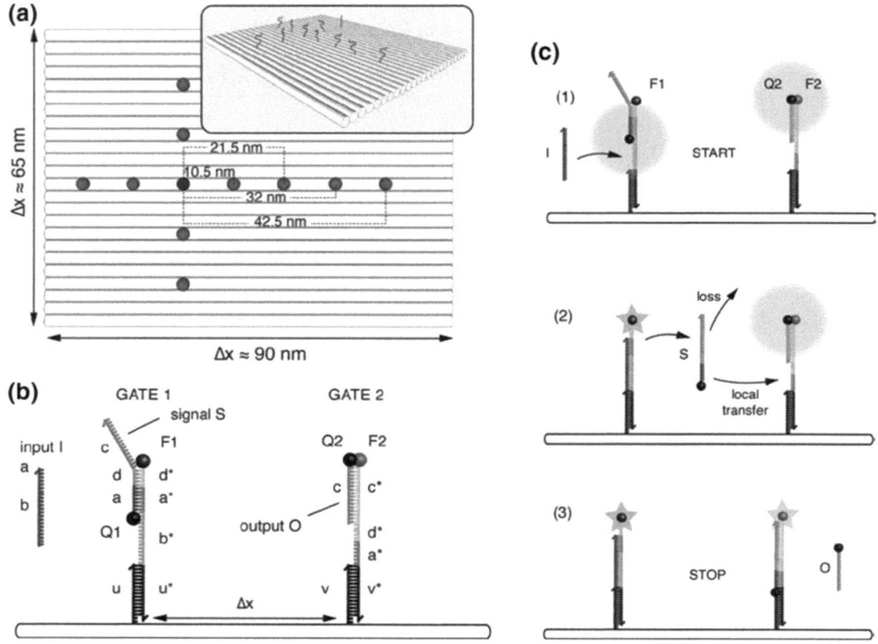

Fig. 2.14 **a** DNA origami was used localize the strand displacement cascade at different separation distances. Four distances between 10.5 and 42.5 nm were investigated. **b** Domain notation of the gate designs. Δx denotes the separation distances between the two gates. **c** The cascade involves three key steps: diffusion and binding of input I to Gate 1, local transfer of signal S strand to Gate 2 and the release of output O (Reprinted with permission from Ref. [78]. Copyright © 2014, American Chemical Society)

reaction modules to control the stepwise interactions and minimize crosstalk between different components [72–75]. The idea of localizing hybridization circuits on a DNA origami substrate was first implemented by Harish et al. [76], where arbitrary Boolean functions are computed by their molecular DNA circuit.

However, localized circuit is a double-edged sword where the increased local concentration promotes the kinetics of both desired and undesired pathways to at least one order of magnitude faster than bulk reaction as long as the DNA components can come into direct contact [77]. The balancing point between the two pathways was investigated by Simmel and coworkers [78], who designed a simple strand displacement cascade on DNA origami with different separation distances between the gates (Fig. 2.14). An optimized inter-gate distance, which in this case was approximately 20 nm, is one way of improving the strand displacement kinetics without excessive leakages. The local reaction was robust against high concentration of competitor strands (up to 3.7 μM), designed to block the active domains (d^* a^*) of triggered signal S from binding to Gate 2.

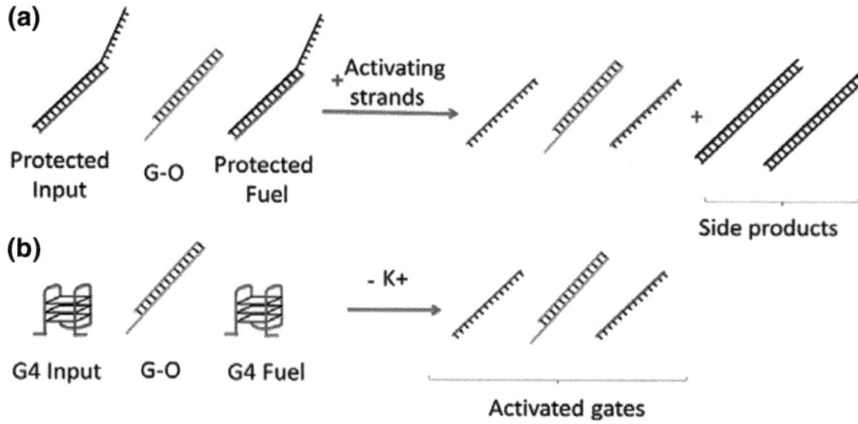

Fig. 2.15 G-quadruplex is used to control the activation of the localized circuit **b** instead of the usual protection and deprotection strand **a** (Reprinted with permission from Ref. [80]. Copyright © 2016, American Chemical Society)

As with other heterogeneous reactions, the surface density needs to be carefully engineered to avoid steric hindrance [79]. More significantly, the implementation of localized circuit increases design complexity which in itself introduces new problems—localizing the circuit speeds up both the desired and leak (direct, non-diffusion limited interactions) reactions; to counter the leak, a set of protection and deprotection strand is needed; and finally the deprotection strand can interact with other spatially separated parts of the circuit which nullifies the compartmentalization advantage from circuit localization.

To avoid using additional protection and deprotection strands, Elezgaray and co-workers [80] proposed to use G-quadruplex DNA to control the activation of the localized DNA circuit which is modulated by environmental changes, e.g. K^+ and NH_4^+ concentration (Fig. 2.15). Unfortunately, the response time of the circuit to changing salt concentration is in the order of minutes which is slower than the local toehold-mediated kinetics (1 s^{-1}) due to the slow unfolding process of the quadruplex. True spatial compartmentalization is achieved using this activation mechanism; however, the circuit also becomes more sensitive to environmental fluctuations.

The works presented thus far use DNA origami to spatially organize strand displacement cascades with the main advantage of drastically improved kinetics due to the localization effect. While the hardwiring of DNA species at specific location along a substrate is required for molecular computing or applications requiring precise spatial control [81], Simmel and co-workers [78] suggested that an alternative gate design which dynamically assembles or disassembles during the circuit operation can be designed to minimize undesired local leak reaction. This essentially sets up a kinetic pathway where the desired processes, now operating in the localized format, occur at a faster kinetics rate than the undesired processes remaining in the bulk solution.

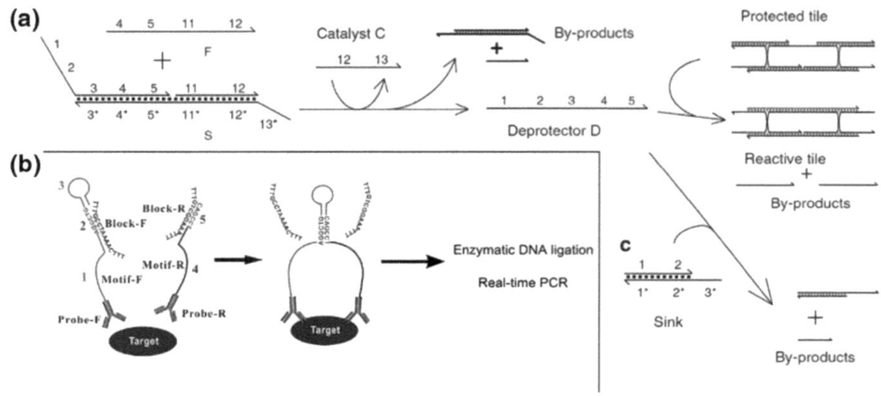

Fig. 2.16 Dynamic assembly of localized strand displacement reactions. **a** Dynamic generation of the catalyst strand (C) to release deprotector strand (D) for formation of reactive origami tiles Ref. [82]. **b** The binding-induced DNA assembly (BINDA) system is used to generate output sequences which can be amplified using PCR to achieve yoctomole detection of protein (Reprinted with permission from Ref. [83]. Copyright © 2012, American Chemical Society)

An interesting idea is to use an upstream circuit reaction to trigger the self-assembly of DNA origami demonstrated by Winfree and co-workers [82], which then forms the substrate for downstream localized circuit operations (Fig. 2.16a). Though, the reaction rate is likely to be limited by the triggering and self-assembly steps. Another possible format is to assemble the circuit spontaneously in response to specific dynamic event(s), e.g. target recognition(s) in biosensing.

The latter concept has been demonstrated by Le and co-workers over several design iterations, all of which are based on the concept of increased local DNA concentration upon target binding. In their binding-induced DNA assembly (BINDA) design, the general concept is to utilize the increased local concentration to promote the interaction kinetics between two initiator strands when bound onto a single protein target to trigger downstream reactions, e.g. polymerase chain reaction (PCR), toehold-mediated strand displacement and CHA (Fig. 2.16b) [83–87].

2.4 Biomedical Applications of DNA Circuit

One major direction of DNA circuit application is for bio-detection and imaging. We will review target systems of increasing complexities, from DNA/RNA to protein and finally whole cell level. More sophisticated operations, such as the use of "smart" DNA nanodevice designs to effect functional output and detect dynamic interaction events, will also be discussed.

The performance of any detection strategy is generally evaluated based on its (1) sensitivity, (2) specificity, (3) speed and (4) ease of operation. One advantage of DNA circuit is the availability of enzyme-free signal amplification designs (Sect. 2.3.3) to boost the signal and hence sensitivity, though the issue of leakage contributing to background cannot be neglected (further discussed in Sect. 2.5.2). The fastest detection speed is limited by the strand displacement rate (approximately 10^6 M^{-1} s^{-1}) which translates to a reaction time scale of seconds. However, the circuit often includes other circuit modules or additional non-DNA based moieties which slows down the overall operation. Pure DNA circuits are often designed to operate autonomously with minimal user intervention, though this simplicity may be traded-off by adding other complex steps to improve other detection metrics, especially sensitivity.

2.4.1 Detection of Nucleic Acids Based Targets

An obvious target for DNA circuit detection is nucleic acids (DNA and RNA) as simple base pairing recognition suffices in triggering the circuit operation which often involves enzyme-free, isothermal signal amplification modules (Sect. 2.3.3). The readout monomer typically has a hairpin stem-loop structure to sequester the toehold domain which is released only upon target binding. This is reminiscent of the classical molecular beacon design [88], which is one of the most widely-studied DNA probe system to date. One improvement offered by toehold-mediated strand displacement designs is that linear-linear hybridization is kinetically more favourable than the linear-loop hybridization used in molecular beacons designs [89].

CHA is a popular signal amplification strategy which can be used flexibly with multiple readout methods, e.g. fluorophore-quencher, electrochemical and colorimetric readout facilitated by G-quadruplex formation, as demonstrated in the proof-of-concept work by Chen and co-workers (Fig. 2.17) [90]. Generally, 20- to 50-fold amplification was achieved per hour and the leakage rate can be reduced from 100 M^{-1} s^{-1} to approximately <0.5 M^{-1} s^{-1} with improved domain design.

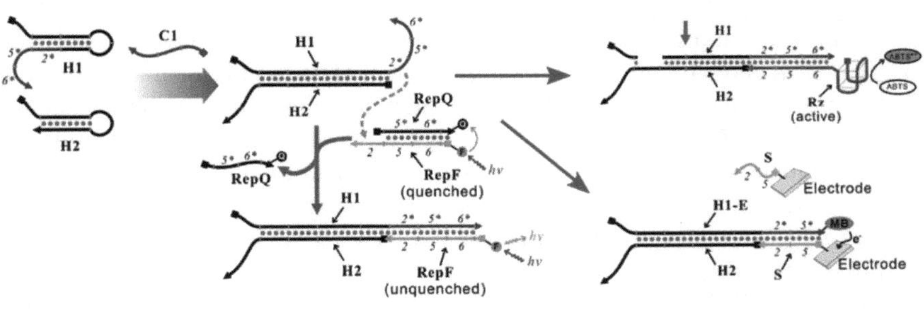

Fig. 2.17 Catalytic hairpin assembly (CHA) for nucleic acids detection, which can be used with different readouts, including fluorescence, colorimetric and electrochemical (Reproduced with permission from Ref. [90]. Copyright © 2011, Oxford University Press)

The limit of detection ranged from 10 pM (electrochemical) to 1.0 nM (colorimetric) over 1–4 h reaction time range.

Since then, various groups have attempted to utilize CHA for sensitive nucleic acids detection. The hairpins can be modified as molecular beacons which lights up during CHA for DNA detection down to 10 pM within 2 h [91]. A lower LOD of 1.0 pM was achieved for RNA detection within the same incubation time, though with much higher salt concentration ($[Na^+] = 600$ nM) [92]. The LOD was pushed down further to 0.3 aM by incorporating gold nanoparticle into the electrochemical readout [93]. However, the overall detection procedure took approximately 5–6 h, with the CHA taking an unsurprisingly long time of 4 h. Ellington and co-workers [94] pushed the detection limit further by stacking two-by-two layers of CHAs to achieve 600,000-fold signal amplification though at an even slower speed of 12 h.

CHA consistently suffers from slow reaction kinetics. To overcome this limitation, Milligan and Ellington [95] demonstrated the use of RecA, a protein that binds with ssDNA and facilitates strand exchange, to improve the catalytic rate up to nine-fold. This reduced the CHA reaction time to less than an hour. Another refreshing idea is to use the initial CHA reaction rate to quantify the DNA detected instead of the absolute readout signal [96]. A reasonable LOD of 0.1 nM was obtained from the kinetics data collected in the first 10 min. The reaction can also be accelerated by adding ethanol to the reaction buffer to detect 9.5 pM of hepatitis B virus DNA within an hour from the original 12 h [97].

Due to the modularity of CHA design, it can be used with existing DNA detection techniques as a signal transducer for further signal amplification. For example, it can be used to generate a simple readout for the loop-mediated isothermal amplification reaction [98], rolling circle amplification (RCA) and strand displacement amplification [99]. Impressive detection limits down to 100 fM–1 pM was achieved. In particular, a general set of "four-rule" principle was established for the effective use of CHA at higher temperature without appreciable leakage, which can be extended to other systems [99]. An even lower LOD of 4.0 aM was achieved by adding an additional step of target-triggered enhanced hybridization process, which demonstrates the scalability of this multi-stacking strategy [100].

Compared to CHA, signal amplification using HCR is faster where most detections are completed within an hour. This could be due to the design of the domain structure. In CHA, the toehold for opening **HP2** is the stem of **HP1** and the toehold for opening **HP1** represents the length by which the catalytic strand needs to spontaneously dissociate from. As such, it is not so straightforward to improve the kinetics by increasing toehold lengths which could lock up the hairpin monomers more tightly or reduce the catalyst regeneration rate. On the other hand, it is clear-cut in HCR that the toehold and loop lengths drive the reaction forward while the stem maintains the hairpin metastability.

However, HCR is a linear amplification scheme which cannot be readily adapted into downstream circuit modules and is typically used as an end-point readout signal. Even so, the question of how to elicit signals from DNA polymerization remains and most works focus on the design of an appropriate readout. For example, G-quadruplex sequences can be included into the hairpin design to develop

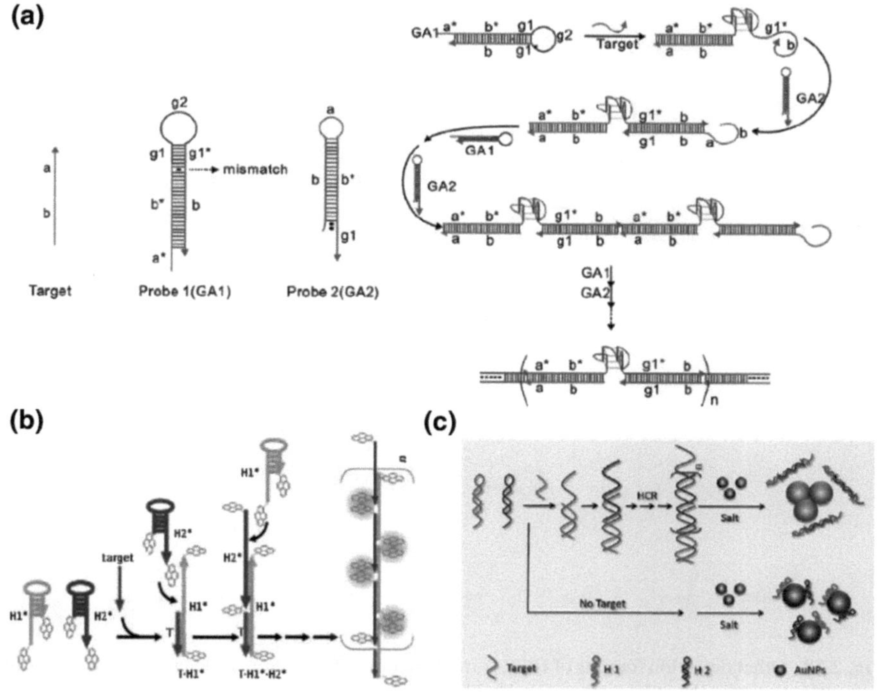

Fig. 2.18 Examples of readout signals that can be incorporated into hybridization chain reaction (HCR): **a** G-quadruplex facilitated colorimetric or fluorometric signal (Reproduced with permission from Ref. [101]. Copyright © 2012 Elsevier B.V.), **b** pyrene excimer formation (Reproduced with permission from Ref. [102]. Copyright © 2010, WILEY-VCH Verlag GmbH & Co. KGaA, Weinheim), **c** aggregation of gold nanoparticles under salt screening conditions (Reprinted with permission from Ref. [105]. Copyright © 2013, American Chemical Society)

colorimetric (LOD = 7.5 nM) and fluorometric (LOD = 4.0 nM) readout (Fig. 2.18a) [101]. A low LOD in the femtomolar range was achieved by modifying the HCR hairpins with pyrene molecule which comes into proximity in the HCR assembly to form fluorescing pyrene excimers (Fig. 2.18b) [102]. Small silver nanocluster can be incorporated for the detection of RNA in both turn-on (LOD = 10 pM) [103] and turn-off (LOD = 0.78 nM) [104] formats. Simple colorimetric methods based on the balance between aggregation of gold nanoparticles under salt screening condition and the electrostatic protection offered by HCR monomers [105] or products [106] are used to achieve sub-nanomolar detection limit (Fig. 2.18c).

The LODs for homogeneous HCR assays are typically in the high picomolar range, which is lower than for CHA assays. The LOD can be improved by using heterogeneous detection where the target strand is captured on a substrate with intermediate washing steps. A LOD of 15 fM was achieved with sequential steps of target binding, HCR amplification and $Ru(phen)_3^{2+}$ electrochemical signal development

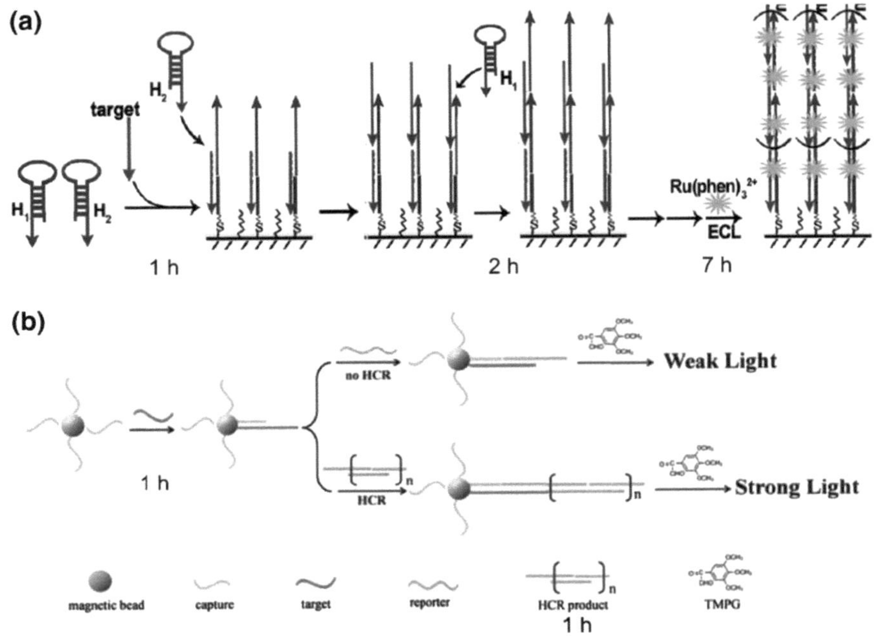

Fig. 2.19 Other detection formats of HCR to improve limit of detection. **a** Heterogeneous detection with electrochemical readout (Reprinted with permission from Ref. [107]. Copyright © 2012, American Chemical Society), **b** Isolation of target strand with magnetic bead for chemiluminescence readout (Reproduced from Ref. [108] with permission from the Royal Society of Chemistry)

(Fig. 2.19a) [107]. However, a longer time (2 h) was needed for the HCR step due to mass transfer limitation while the total assay time was 10 h.

Alternatively, magnetic beads can be used to isolate the target strands for selective amplification (Fig. 2.19b). Typically, the LODs are improved to femtomolar or low picomolar range by including separation step(s) and readouts, e.g. chemiluminescence [108], enzymatic amplification [109] and enzyme-induced silver metallization [110]. Since these detection schemes operate in homogeneous phase, the HCR reactions can still be completed within an hour. However, additional processing steps, e.g. capture of target strands and development of readout signals, are required which increase the assay time to about 3 h.

2.4.2 *Detection of Other Biomolecular Targets*

Other biomolecular targets, e.g. proteins and small molecules, can be detected by incorporating the appropriate recognition moiety, e.g. aptamers and antibodies, or designing a translator module (Sect. 2.3.2) to convert the recognition events into

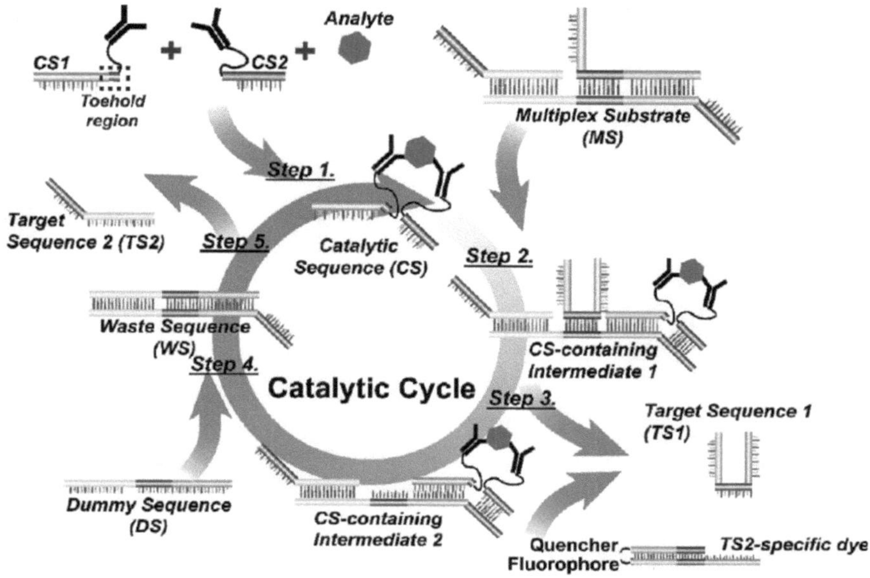

Fig. 2.20 Antibody-DNA probes were used for detecting the target protein analyte and trigger the entropy-driven catalytic cycle (Reprinted with permission from Ref. [116]. Copyright © 2016, American Chemical Society)

DNA-based readouts [111–114]. General designs with the potential to go beyond proof-of-concept models will be highlighted.

The use of dual biomolecular recognition events appears to be a recurring theme in these designs. On the front of aptamer-based detection, Ju and co-workers [115] split a single aptamer strand and incorporated them into two recognition probes. Upon target binding, the two probe are brought into proximity for signal transduction into CHA readout. The circuit can operate AND, INHIBIT and NAND gates for different input/output combinations. A wider pool of targets can be recognized by this split aptamer strategy rather than methods requiring a pair of complete aptamers for binding to a single target, often thrombin. However, the detection speed of approximately 2–4 h still needs further improvements.

The golden standard for protein detection, which represents a huge pool of existing biomarkers, is based on antibody-based recognition. Di Carlo and co-workers [116] recently designed an entropy-driven circuit where the dual recognition by a pair of antibody-DNA probes triggers an autonomous catalytic cycle for amplified detection (Fig. 2.20). This homogeneous assay requires only one addition step under isothermal condition to achieve LOD in the femtomolar range within 10 min of operation. The LOD can be further pushed down to detect 4 aM of influenza A nucleoprotein using a digital format in microfluidic well array. The short detection time is likely due to the faster entropy-driven strand displacement reaction as opposed to the more commonly used hairpin-mediated amplification.

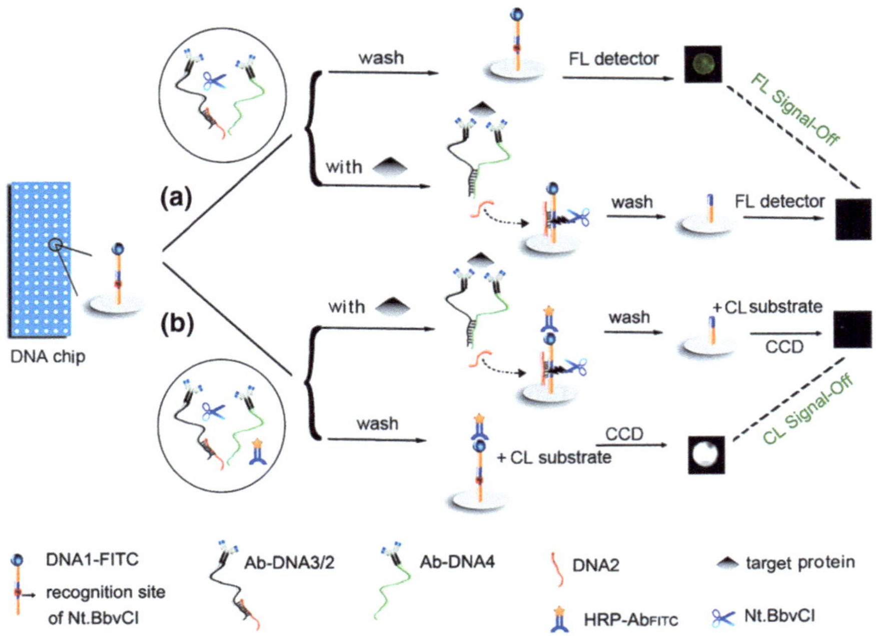

Fig. 2.21 A DNA array platform for high throughput analysis of multiple protein targets through a series of strand displacement and endonuclease cleavage reactions to generate turn-off **a** fluorescence and **b** chemiluminescence signal. (Reproduced from Ref. [117] under the Creative Commons Attribution 3.0 Unported License)

Ju and co-workers [117] developed a high throughput array platform for the simultaneous detection of 4 cancer biomarkers in 5 patient serum samples within 35 min of incubation and with a detection error of less than 8.7%. This technology used a pair of antibody-DNA probes to recognize the target protein(s) (Fig. 2.21). One DNA probe is hybridized with a short (11 nt) protector strand which is displaced by the complementary domain on the second DNA probe upon target binding. The short protector strand then transduces the recognition signal by binding to an immobilized DNA labelled with fluorophore dye. A nicking endonuclease cleaves the double stranded region and releases the dye which is then washed away to give a turn-off signal. Low LODs in the pg mL^{-1} range was achieved, which can possibly be improved by incorporating a turn-on signal instead.

2.4.3 Cell Imaging and Visualization

DNA circuits have been developed for cell-based detection and visualization, often with simple designs for practical applicability. HCR is the leading concept for the in situ mapping of expression levels as the linear, amplified DNA chain remains tethered to the target recognized.

Pierce and co-workers [118] successfully designed five sets of orthogonal HCR hairpins for the multiplexed imaging of mRNA expression [119] which was previously a challenging task due to cross-talks between detection methods (Fig. 2.22a). This is made possible by the programmability and specificity of HCR for simultaneous operations, which sequences can later be evolved to improve the signal gain [120]. A wash-free version was recently developed by incorporating FRET readout into HCR for imaging tumour-related mRNA [121].

Nonetheless, simultaneous detection is limited by the number of spectrally distinct fluorophores available. One way is to use erasable DNA complexes to image different cellular targets within the same cell sequentially (Fig. 2.22b) [122]. The toehold overhang in the labelling strand allows the current fluorophore to be removed via strand displacement for the next set of probes to be used on other targets. A total of six proteins were detected using antibody-DNA conjugates. The strand removal process is complete with minimal perturbation to the sample.

While the above studies were performed on fixed cells, several circuit designs have matured sufficiently for proof-of-concept imaging in live cells. An early demonstration by Hemphill and Deiters [123] suggests that DNA circuits can be delivered into live cells using common transfection agent to perform a simple AND gate operation within the cellular environment. It is crucial for circuits to operate as one-pot in live cell imaging since washing steps cannot be incorporated. For example, a fluorophore-quencher readout can be used with CHA for the amplified detection of mRNA in live cells [124]. However, precise spatial information cannot be obtained from the non-localized CHA readout.

HCR can be used as a localized, turn-on amplifier by incorporating FRET readout. This was demonstrated by Jiang and co-workers [125] who used gold nanoparticle to deliver labelled HCR hairpin strands into live cells, and Weizmann and co-workers [126] who modified the domain arrangements of HCR to accommodate FRET readout, in a scheme termed cascade hybridization reaction, for miRNA imaging. However, the circuit kinetics was slowed down considerably in these live cell works and the detection took about 3–4 h in total. This could be due to the relatively low salt and crowded environment within the cells [127] and further improvements to the circuit design is needed to boost the reaction rate without incurring significant background leakage.

Besides performing detection within the cells, circuits can be used to evaluate cell surface markers. This application may be more suitable for synthetic DNA devices since the accessibility of cell surface markers obviates the need for transfection. Tan and co-workers [128] first demonstrated that several types of fluorescent DNA nanodevices, e.g. self-assembled DNA chain using HCR, can be built on live cell surfaces for imaging on confocal microscope and flow cytometer (Fig. 2.23a). Besides imaging specific cell surface proteins for end-point detection, Stojanovic and co-workers [129] showed that simple logic gate computations can be executed using strand displacement cascade to profile the cell surface markers and classify specific

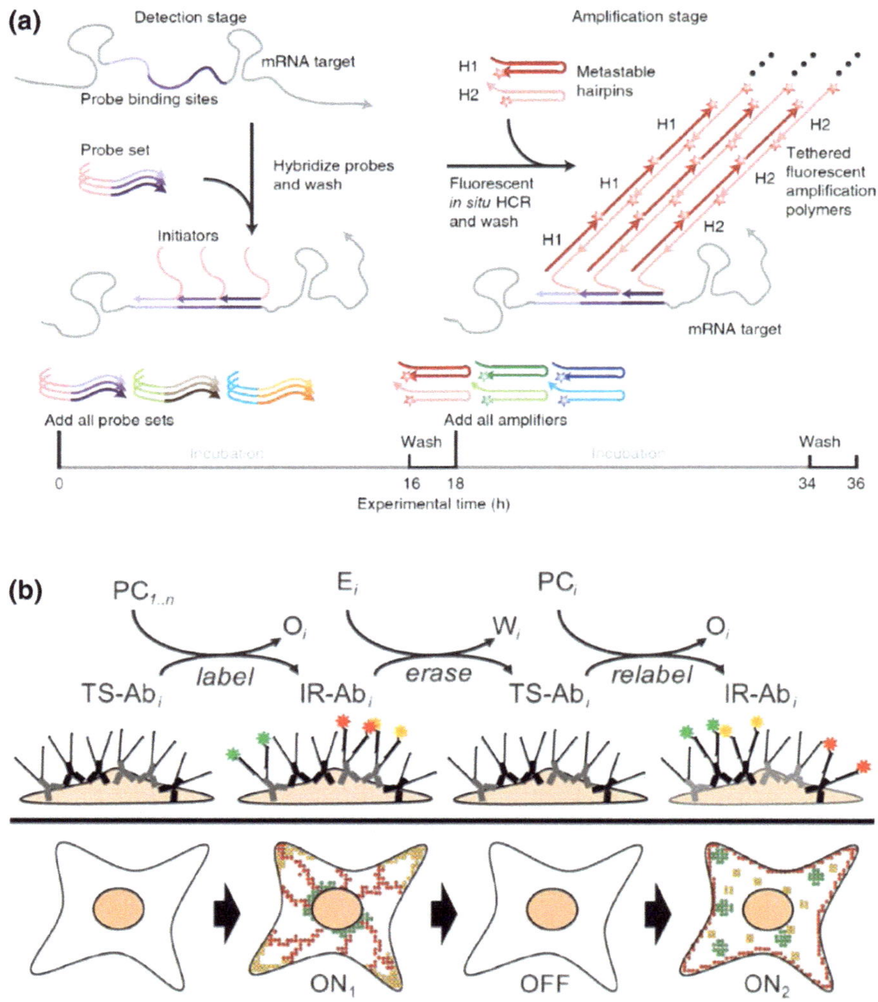

Fig. 2.22 Circuit designs for multiplexed mRNA imaging in cell. **a** Multiple mRNA can be imaged simultaneously using HCR hairpins with orthogonal sequence designs (Reproduced with permission from Ref. [118]. Copyright © 2010, Springer Nature), **b** Sequential imaging was performed using erasable DNA complexes (Reproduced with permission from Ref. [122]. Copyright © 2012, WILEY-VCH Verlag GmbH & Co. KGaA, Weinheim)

subpopulation of lymphocytes (Fig. 2.23b). This is a step forward towards the design of "smart" devices that leverages on the computation capability of DNA circuits.

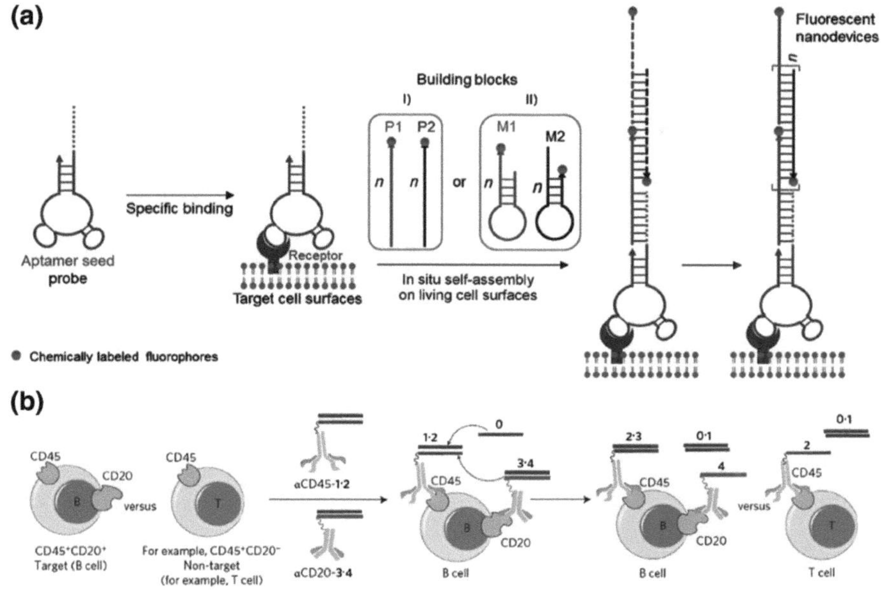

Fig. 2.23 DNA circuits can be used to detect cell surface proteins *via* **a** hybridization chain reaction (HCR) for visualization (Reproduced with permission from Ref. [128]. Copyright © 2013, WILEY-VCH Verlag GmbH & Co. KGaA, Weinheim) and **b** strand displacement cascade for evaluating the cell surface marker profile (Reproduced with permission from Ref. [129]. Copyright © 2013, Springer Nature)

2.4.4 "Smart" DNA Nanodevices

Since DNA circuits can perform autonomous computation, it is useful if the output generated has a functional effect besides as an end-point readout signal. For example, Tan and co-workers [130] demonstrated the regulation of thrombin activity based on aptamer recognition which triggers a circuit operation to release an inhibitor strand (also an aptamer) if the thrombin concentration exceeds a pre-set threshold level (Fig. 2.24a). This feedback loop is a neat demonstration which mimics the homeostasis processes in biological systems, though the design needs to be expanded beyond targets with multiple, well-characterised aptamer-response moieties to find general utility.

Another direction for "smart" detection is in theranostics, where dual functions of therapy and diagnostics are executed in a single platform. For example, Tan and co-workers [131] used aptamer moiety to anchor a trigger strand onto cells expressing specific surface marker to initiate CHA for amplified photodynamic effect thereby killing the cancer cells. Again based on aptamer recognition, Chu and co-workers [132] used HCR hairpins, one carrying fluorophore-quencher readout and the other carrying a prodrug, to deliver the theranostic agents into specific tumour cells via

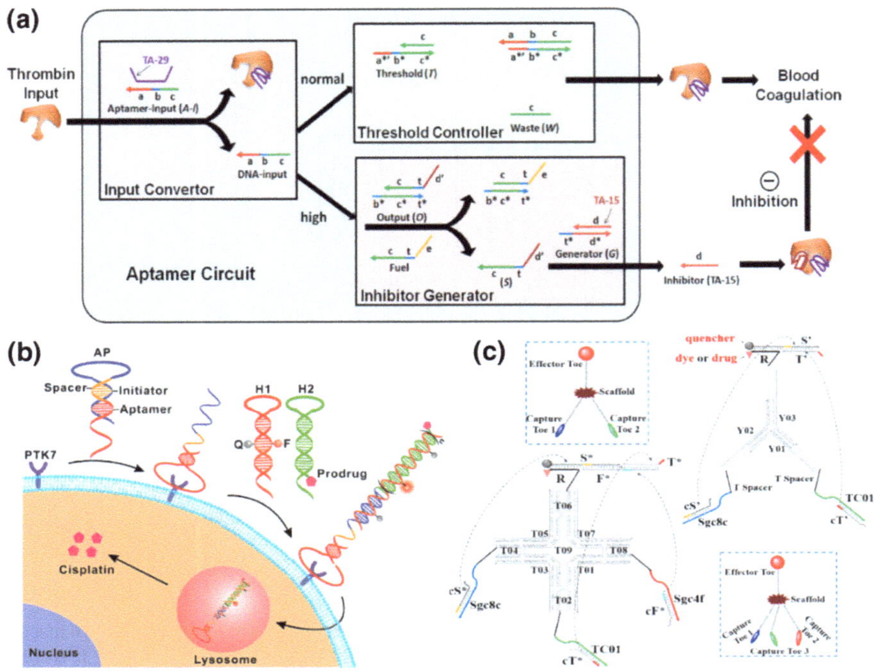

Fig. 2.24 Designs of "smart" DNA nanodevices capable of **a** performing regulatory feedback loops (Reprinted with permission from Ref. [130]. Copyright © 2012, American Chemical Society), **b** inducing theranostics effect on specific tumour cells (Reprinted with permission from Ref. [132]. Copyright © 2015, American Chemical Society) and **c** multi-target evaluation for targeted therapy (Reprinted with permission from Ref. [134]. Copyright © 2014, American Chemical Society)

endocytosis for selective cell imaging and killing (Fig. 2.24b). Wang and co-workers [133] recently reported a split aptamer-based activatable theranostic probe (SATP) where a pair of split aptamer arms reconfigures in response to the recognition of target proteins. This displaces a short quencher strand to activate a turn-on fluorescence signal and the release of intercalated doxorubicin.

The gate design(s) can be scaled up for recognizing multiple biomolecular targets. For example, Tan and co-workers [134] designed a DNA "nanoclaw" where the aptamer-based recognition claws are held together with the effector moiety to minimize cross-talks between individual nanoclaw and maximize the output (drug or imaging dye) released due to their close proximity (Fig. 2.24c). Complex multiparametric computation (12 three-input conditions and 2 four-input conditions) can be executed through careful gate designs to evaluate a wider range of cell types and dispense the appropriate therapeutic reagents; almost like a little clinic in the body [135].

A recurring theme in the above designs is the use of aptamer as the recognition moiety. Once the pool of well-validated aptamer sequences expands, we expect this direction of "smart" cell-based computation for precision medicine applications to grow [136].

2.4.5 Detection of Biomolecular Interactions

Biomolecular interactions are dynamic events underlying numerous biological pathways and cellular functions, e.g. cell receptor clustering for signal transduction [137], host-cell interaction during infections [138] and within protein complexes [139]. Despite the biological significance, a simple molecular toolbox for the direct detection of such multi-target events is lacking [140] even though a myriad of tools are available for the end-point detection of individual biomolecules [141–143]. Furthermore, biomolecular interactions exist in various forms—stable or transient complex, with strong or weak binding, which can further complicate analysis [144]. We will provide brief background information on conventional detection techniques and later, introduce existing DNA-based detection assays.

The most common analysis methods are co-immunoprecipitation and pull-down assay [145]. They require the use of a "bait" (antibody for the former, and any protein of interest for the latter) to fish for the "prey", or its interacting partner, from cell lysate. The use of an artificial "bait" immobilized on a bead only captures the relative bulk expression levels under static condition. Moreover, extensive sample processing steps are needed to isolate biomolecules from their original environment which is cumbersome and may destroy their structures. In vivo methods involving genetic techniques, e.g. yeast two-hybrid system [146] and protein fragment complementation assay [147], offer a more representative picture of the interacting complexes under cellular context [148]. The general concept is that the "baits" are fused separately with two proteins of interests ("prey") and when brought into proximity by an interaction event, stimulates or refolds into a detectable output. Though the detection occurs in cells, the protein components are fused together unnaturally and a positive indication of interaction need not reflect biological relevance. As such, they are useful for the large-scale screening of interaction partners, but not for in situ analysis of actual complexes.

A DNA-based, in situ detection method called proximity ligation assay (PLA) was developed in recent years to interrogate biomolecular interactions over longer long distance range beyond the FRET limit (2–10 nm). PLA involves a pair of proximity probes coming into proximity when bound to the target and hybridize with a connector oligonucleotide for enzymatic ligation [149]. The ligated oligonucleotides are used to trigger downstream readouts, e.g. PCR and RCA (Fig. 2.25a) [150]. However, the amplified signals tend to saturate easily, partly due to the depletion of substrates, and so this method is more suitable for analyzing targets of low abundance [151]. The dynamic range of RCA-based PLA can be improved by compacting the RCA product [152] or by using a combination of different fluorophore-labelled detection oligonucleotides [153].

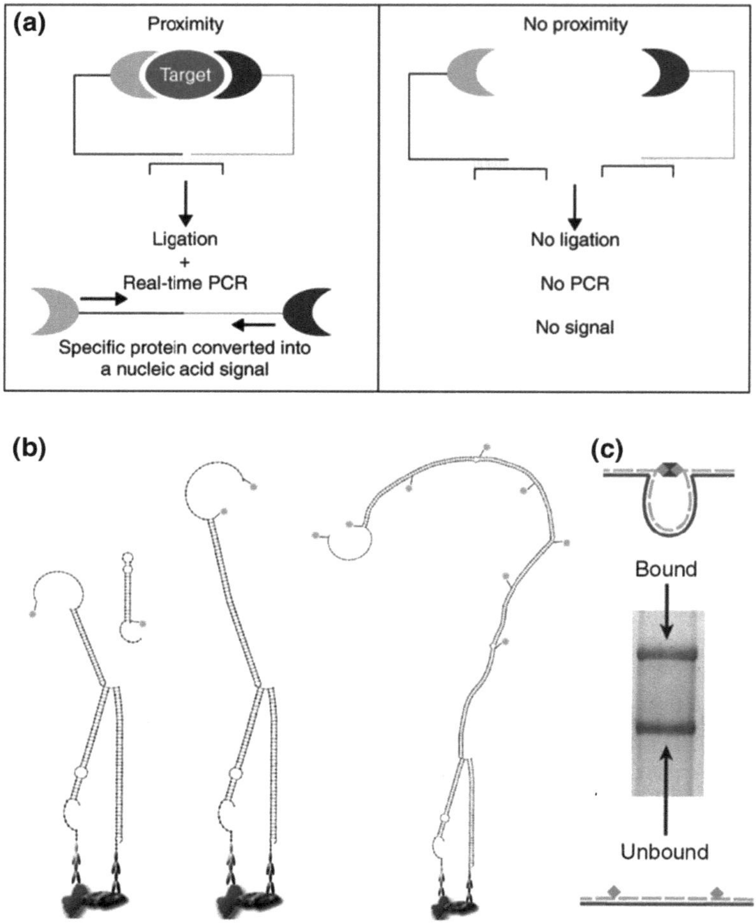

Fig. 2.25 DNA-based methods for detecting biomolecular interactions. They include **a** proximity ligation assay (PLA) with PCR readout (Reproduced with permission from Taylor & Francis, Ref. [150]) or **b** hybridization chain reaction (HCR) readout, Ref. [154] and **c** DNA nanoswitch (Reproduced with permission from Ref. [155]. Copyright © 2014, Springer Nature)

Interestingly, an enzyme-free variant was recently developed based on the proximity-dependent initiation of HCR (Fig. 2.25b) [154]. Still, the execution requires a series of probe addition and washing steps. This is because the generation of signals is not triggered exclusively by proximal events and so it is necessary to remove unbound probes and unreacted reagents before meaningful results can be obtained.

The methods reviewed thus far suffer from a common issue of lengthy and cumbersome experimental steps. A simple yet effective DNA-based technique, called DNA nanoswitch, was recently developed by Wong and co-workers [155]. It is a simple, low-cost and label-free method for interrogating biomolecular interactions and detecting nucleic acid sequences [156] based on conformation changes in the DNA structure (Fig. 2.25c). Though a versatile and powerful tool requiring only gel electrophoresis for readout, it requires the complexes of interest to be removed from its original state and cannot be used to perform in situ analysis and visualization. Besides, migration through gel electrophoresis is likely to disrupt weak interaction events. As such, a simple toolbox which can simultaneously perform both one-pot and in situ detection of biomolecular interaction events is desirable.

2.5 Current Challenges Faced in DNA Circuit

2.5.1 Unidirectional Reaction Pathway

DNA circuit is driven by the free energy of DNA hybridization to fuel the strand displacement cascade. Hence, the reaction stops once the DNA reactants are depleted [27]. A common method to regenerate the DNA devices is to use another DNA strand (typically called fuel strand); however, this merely shifts the limiting reagent to another species which does not solve the root of the problem, i.e. the lack of a continuous energy source [2]. More significantly, the circuit operation is highly dependent on the initial input condition and cannot truly respond real-time to changes in the environment when the fixed pool of DNA strands is depleted. It will be interesting if self-sustaining, reversible circuit can be designed in the future to process real-time changes in the input parameter(s). A short-term fill-gap solution is to input sufficiently high DNA concentration at the start to sustain the operation for a meaningful time frame.

2.5.2 Circuit Leakage

Using too high a DNA concentration, however, exacerbates another issue of circuit leakage which affects the performance of almost all circuit designs. There are three main sources of leakages—impurities arising from synthesis defects, toehold-independent strand displacement and cross-talk.

First, oligonucleotides used in most studies are obtained via solid-state synthesis with approximately 1% failure rate in each coupling step to generate a mixture of complete and defected strands [157]. The defected strands can truncate the desired reaction or promote undesired pathways [158]. We will revisit this issue in Chap. 6. Next, toehold-independent strand displacement is facilitated by the "breathing" of DNA helices at its ends. "Breathing" is a consequence of the thermal fluctuation within the DNA where hydrogen bonds between base pairs are disrupted momentarily and the bases flip out of the helical structure [159]. At room temperature, the base

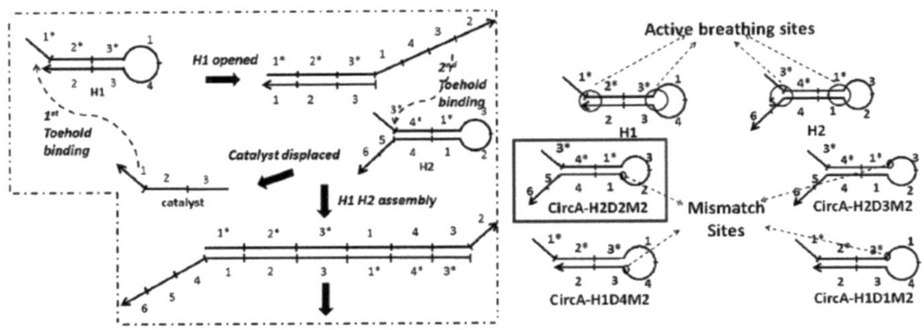

Fig. 2.26 Using base pair mismatch to reduce leakages in catalytic hairpin amplification (CHA) (Reproduced with permission from Ref. [162]. Copyright © 2014, WILEY-VCH Verlag GmbH & Co. KGaA, Weinheim)

pairs at the termini are 50% open, and the next in line are 10–20% open while the remaining base pairs have much lower opening probability of approximately 10^{-6} [160]. A·T base pairs are weaker and is more likely to "breathe", so Zhang [161] recommends that the branch migration domain start and end with guanine or cytosine nucleotide.

Strategies to counter breathing includes the use of "clamp" sequences [65] and by introducing mismatch at selected sites. The former was found to be an ineffective method when excessive lengths were used [90]. Ellington and co-workers [162] demonstrated that the introduction of mismatches at the 3′-end of **HP2** domain *2* is most effective in supressing CHA leakages since it counteracts "breathing" at the termini of **HP1** stem where domain *2** is (Fig. 2.26). We will further argue in Chapter 6 that having mismatch at this position also helps to supress leakages due to synthesis defect.

2.5.3 Sensitivity to Salt Condition

DNA circuit operation is sensitive to the salt concentration in solution and high Na^+ or Mg^{2+} concentration is often needed to promote the circuit kinetics. Circuit operations using physiological concentration of 140 mM Na^+ are typically compensated by increasing Mg^{2+} up to 12.5 mM (note that 1 M Na^+ is thermodynamically equivalent to 12.5 mM Mg^{2+}) [17]. Some of the typical reaction conditions are presented in Table 2.1. Circuits tested under physiological conditions remains as a minority [90, 129]. Most systems can only work under low Na^+ concentration with Mg^{2+} supplementation. CHA is again revealed as a slow signal amplification strategy even when 12.5 mM of Mg^{2+} was used [124]. HCR is a popular strategy for visualization at cellular level. The basic toehold-mediated strand displacement reaction stands out for its fast operation under physiological condition—this may be the most feasible design for live cell works if signal amplification is not necessary.

Table 2.1 Reaction conditions of selected circuits for detection applications

Application	Description	Reaction conditions			Reaction time	References
		NaCl (mM)	$MgCl_2$ (mM)	Other buffer components		
Biomolecular detection	Hybridization chain reaction (HCR) for DNA detection	500	0	50 mM H_2PO_4 (pH 6.8)	N/A	[62]
	HCR for aptamer-based detection	300	5	20 mM Tris (pH 7.6)	N/A	
	HCR for detecting DNA targets in biological environment	1000	–	Normal: 1X SPSC Biological: DMEM + 10% FBS	1 h	[102]
	Catalytic hairpin amplification (CHA) used with different readout for DNA and ATP detection*	140	0	5 mM KCl, 20 mM Tris (pH 7.5)	1–4 h	[90]
	Combination of CHA with associative toehold for protein detection	140	10	PBS-Mg buffer + 0.05% Tween20 at 37 °C	2.5 h	[84]
	Associative toehold for DNA detection	140	5	20 mM Tris (pH 7.5), 5 mM KCl, 0.5 mM EDTA	N/A	[53]
	Associative toehold with polymerase chain reaction (PCR) readout for protein detection	–	10	TE-Mg buffer + 0.05% T-20	30 min	[83]
Fixed cell imaging	Multiplexed in situ immunofluorescence of immune cell surface markers using erasable DNA complexes	140	12.5	2% BSA, 1 mg/mL herring sperm DNA, 0.5 μM PolyT DNA in PBS	Signal development: 2 h Probe erasing: overnight incubation (*ca.* 12 h)	[122]

(continued)

Table 2.1 (continued)

Application	Description	Reaction conditions			Reaction time	References
		NaCl (mM)	$MgCl_2$ (mM)	Other buffer components		
	HCR for multiplexed imaging mRNA in fixed cells	750	–	5X SSCT and other blocking agents	Overnight incubation (12–16 h)	[120]
	Proximity ligation assay (PLA) using HCR readout	1000	–	50 mM Na_2HPO_4 (pH 7.4)	10 min sufficient; equilibrium 30 min	[154]
	Associative toehold for cell imaging	140	10	PBS-Mg buffer, 0.05% Tween20	5 min	[163]
Live cell imaging	DNA strand displacement and HCR with aptamer recognition for cell imaging	140	5.0	DBPS supplemented with $MgCl_2$	30 min	[128]
	DNA "nanoclaw" design to autonomously compute multiple cell surface marker inputs for detection and photodynamic therapy	140	5.0	DBPS supplemented with $MgCl_2$	30 min	[134]
	Strand displacement on cell surface*	140	1.2	1X PBS + $MgCl_2$	*ca.* 15–20 min	[129]
	mRNA imaging in live cell using CHA	–	12.5	TAE-Mg (pH 8.0)	2-4 h	[124]
	mRNA imaging in live cell using HCR in a see-saw design	140	5	1X PBS (in vitro) DMEM + 10% FBS (in vivo) 37 °C	4 h	[126]

*Denotes works utilizing physiologically relevant salt condition

2.6 Concluding Remarks

DNA circuit has matured from the simple concept of toehold-mediated strand displacement to more sophisticated toehold and gate designs. On the biomedical front, numerous detection strategies have been demonstrated across different levels of biomolecules. Two concepts stand out—first, the enzyme-free nature of circuit amplification schemes and next, the ability to translate non-nucleic acids target into DNA outputs for circuit activation. Yet, few designs are translated into practical applications. One reason is the recurring issue of circuit leakages which is still not well understood. The use of localized DNA circuit to partly negate the effect of leakage in the bulk solution is an interesting proposition which can be generally applied across different design concepts.

References

1. Regalado, A. MIT Technology Review. https://www.technologyreview.com/s/400727/dna-computing/ (2000)
2. Chen, X., Ellington, A.D.: Shaping up nucleic acid computation. Curr. Opin. Biotechnol. **21**, 392 (2010)
3. Watson, J.D., Crick, F.H.C.: Molecular structure of nucleic acids. Nature **171**, 737 (1953)
4. Yakovchuk, P., Protozanova, E., Frank-Kamenetskii, M.D.: Base-stacking and base-pairing contributions into thermal stability of the DNA double helix. Nucleic Acids Res. **34**, 564 (2006)
5. Hunter, C.A.: Sequence-dependent DNA structure. J. Mol. Biol. **230**, 1025 (1993)
6. Karabyk, H., Sevincek, R., Karabyk, H.: π-cooperativity effect on the base stacking interactions in DNA: is there a novel stabilization factor coupled with base pairing H-bonds? Phys. Chem. Chem. Phys. **16**, 15527 (2014)
7. Vallone, P.M., Paner, T.M., Hillario, J., Lane, M.J., Faldasz, B.D., Benight, A.S.: Melting studies of short DNA hairpins: influence of loop sequence and adjoining base pair identity on hairpin thermodynamic stability. Biopolymers **50**, 425 (1999)
8. Rentzeperis, D., Alessi, K., Marky, L.A.: Thermodynamics of DNA hairpins: contribution of loop size to hairpin stability and ethidium binding. Nucleic Acids Res. **21**, 2683 (1993)
9. Senior, M.M., Jones, R.A., Breslauer, K.J.: Influence of loop residues on the relative stabilities of DNA hairpin structures. Proc. Natl. Acad. Sci. U. S. A. **85**, 6242 (1988)
10. Kuznetsov, S.V., Ren, C.-C., Woodson, S.A., Ansari, A.: Loop dependence of the stability and dynamics of nucleic acid hairpins. Nucleic Acids Res. **36**, 1098 (2008)
11. SantaLucia, J., Hicks, D.: The thermodynamics of DNA structural motifs. Annu. Rev. Biophys. Biomol. Struct. **33**, 415 (2004)
12. Kuznetsov, S.V., Ansari, A.: A kinetic zipper model with intrachain interactions applied to nucleic acid hairpin folding kinetics. Biophys. J. **102**, 101 (2012)
13. Nelson, J.W., Tinoco, I.: Comparison of the kinetics of ribo-, deoxyribo- and hybrid oligonucleotide double-strand formation by temperature-jump kinetics. Biochemistry **21**, 5289 (1982)
14. Owczarzy, R., Vallone, P.M., Gallo, F.J., Paner, T.M., Lane, M.J., Benight, A.S.: Predicting sequence-dependent melting stability of short duplex DNA oligomers. Biopolymers **44**, 217 (1997)
15. Delcourt, S.G., Blake, R.D.: Stacking energies in DNA. J. Biol. Chem. **266**, 15160 (1991)
16. Frank-Kamenetskii, M.D.: Simplification of the empirical relationship between melting temperature of DNA, Its GC content and concentration of sodium ions in solution. Biopolymers **10**, 2623 (1971)

17. Owczarzy, R., Moreira, B.G., You, Y., Behlke, M.A., Walder, J.A.: Predicting stability of DNA duplexes in solutions containing magnesium and monovalent cations. Biochemistry **47**, 5336 (2008)
18. Crick, F.: Central dogma of molecular biology. Nature **227**, 561 (1970)
19. Jackson, D.A., Symons, R.H., Berg, P.: Biochemical method for inserting new genetic information into DNA of Simian Virus 40: circular SV40 DNA molecules containing lambda phage genes and the galactose operon of *Escherichia Coli*. Proc. Natl. Acad. Sci. U. S. A. **69**, 2904 (1972)
20. Seeman, N.C.: Nucleic acid junctions and lattices. J. Theor. Biol. **99**, 237 (1982)
21. Seeman, N.C.: Structural DNA Nanotechnology. Cambridge University Press (2016)
22. Lu, Y., Liu, J.: Functional DNA nanotechnology: emerging applications of DNAzymes and aptamers. Curr. Opin. Biotechnol. **17**, 580 (2006)
23. Ellington, A.D., Szostak, J.W.: In vitro selection of RNA molecules that bind specific ligands. Nature **346**, 818 (1990)
24. Ellington, A.D., Szostak, J.W.: Selection in vitro of single-stranded DNA molecules that fold into specific ligand-binding structures. Nature **355**, 850 (1992)
25. Adleman, L.M.: Molecular computation of solutions to combinatorial problems. Science **266**, 1021 (1994)
26. Yurke, B., Turberfield, A.J., Mills, A.P., Simmel, F.C., Neumann, J.L.: A DNA-fuelled molecular machine made of DNA. Nature **406**, 605 (2000)
27. Zhang, D.Y., Seelig, G.: Dynamic DNA nanotechnology using strand-displacement reactions. Nat. Chem. **3**, 103 (2011)
28. Zhang, D.Y., Turberfield, A.J., Yurke, B., Winfree, E.: Engineering entropy-driven reactions and networks catalyzed by DNA. Science **318**, 1121 (2007)
29. Reynaldo, L.P., Vologodskii, A.V., Neri, B.P., Lyamichev, V.I.: The kinetics of oligonucleotide replacements. J. Mol. Biol. **297**, 511 (2000)
30. Yurke, B., Mills, A.P.: Using DNA to power nanostructures. Genet. Program. Evol. M. **4**, 111 (2003)
31. Zhang, D.Y., Winfree, E.: Control of DNA strand Displacement kinetics using toehold exchange. J. Am. Chem. Soc. **131**, 17303 (2009)
32. Panyutin, I.G., Hsieh, P.: The kinetics of spontaneous DNA branch migration. Proc. Natl. Acad. Sci. U. S. A. **91**, 2021 (1994)
33. Srinivas, N., Ouldridge, T.E., Šulc, P., Schaeffer, J.M., Yurke, B., Louis, A.A., Doye, J.P.K., Winfree, E.: On the biophysics and kinetics of toehold-mediated DNA strand displacement. Nucleic Acids Res. **41**, 10641 (2013)
34. Zadeh, J.N., Steenberg, C.D., Bois, J.S., Wolfe, B.R., Pierce, M.B., Khan, A.R., Dirks, R.M., Pierce, N.A.: NUPACK: analysis and design of nucleic acid systems. J. Comput. Chem. **32**, 170 (2011)
35. Owczarzy, R., Tataurov, A.V., Wu, Y., Manthey, J.A., McQuisten, K.A., Almabrazi, H.G., Pedersen, K.F., Lin, Y., Garretson, J., McEntaggart, N.O., Sailor, C.A., Dawson, R.B., Peek, A.S.: IDT SciTools: a suite for analysis and design of nucleic acid oligomers. Nucleic Acids Res. **36**, W163 (2008)
36. Zuker, M.: Mfold web server for nucleic acid folding and hybridization prediction. Nucleic Acids Res. **31**, 3406 (2003)
37. SantaLucia, J.: A unified view of polymer, dumbbell, and oligonucleotide DNA nearest-neighbor thermodynamics. Proc. Natl. Acad. Sci. U. S. A. **95**, 1460 (1998)
38. Lakin, M.R., Youssef, S., Polo, F., Emmott, S., Phillips, A.: Visual DSD: a design and analysis tool for DNA strand displacement systems. Bioinformatics **27**, 3211 (2011)
39. Ouldridge, T.E., Louis, A.A., Doye, J.P.K.: Structural, mechanical, and thermodynamic properties of a coarse-grained DNA model. J. Chem. Phys. **134**, 085101 (2011)
40. Seelig, G., Soloveichik, D., Zhang, D.Y., Winfree, E.: Enzyme-free nucleic acid logic circuits. Science **314**, 1585 (2006)
41. Green, S.J., Lubrich, D., Turberfield, A.J.: DNA hairpins: fuel for autonomous DNA devices. Biophys. J. **91**, 2966 (2006)

42. Genot, A.J., Zhang, D.Y., Bath, J., Turberfield, A.J.: Remote toehold: a mechanism for flexible control of DNA hybridization kinetics. J. Am. Chem. Soc. **133**, 2177 (2011)
43. Zhu, J., Zhang, L., Zhou, Z., Dong, S., Wang, E.: Molecular aptamer beacon tuned DNA strand displacement to transform small molecules into DNA logic outputs. Chem. Commun. **50**, 3321 (2014)
44. Xing, Y., Yang, Z., Liu, D.: A responsive hidden toehold to enable controllable DNA strand displacement reactions. Angew. Chem. Int. Ed. **50**, 11934 (2011)
45. Zhang, L., Guo, S., Zhu, J., Zhou, Z., Li, T., Li, J., Dong, S., Wang, E.: Engineering DNA three-way junction with multifunctional moieties: sensing platform for bioanalysis. Anal. Chem. **87**, 11295 (2015)
46. Feng, S., Wu, F., Xu, J., Chen, Y., Zhou, X.: Regulation of DNA strand displacement using a G-quadruplex-mediated toehold. RSC Adv. **4**, 55367 (2014)
47. Amodio, A., Zhao, B., Porchetta, A., Idili, A., Castronovo, M., Fan, C., Ricci, F.: Rational design of pH-controlled DNA strand displacement. J. Am. Chem. Soc. **136**, 16469 (2014)
48. Deng, W., Xu, H., Ding, W., Liang, H.: DNA logic gate based on metallo-toehold strand displacement. PLoS ONE **9**, e111650 (2014)
49. Tang, W., Wang, H., Wang, D., Zhao, Y., Li, N., Liu, F.: DNA tetraplexes-based toehold activation for controllable DNA strand displacement reactions. J. Am. Chem. Soc. **135**, 13628 (2013)
50. Prokup, A., Hemphill, J., Deiters, A.: DNA computation: a photochemically controlled and gate. J. Am. Chem. Soc. **134**, 3810 (2012)
51. Prokup, A., Hemphill, J., Liu, Q., Deiters, A.: Optically controlled signal amplification for DNA computation. ACS Synth. Biol. **4**, 1064 (2015)
52. Huang, F., You, M., Han, D., Xiong, X., Liang, H., Tan, W.: DNA branch migration reactions through photocontrollable toehold formation. J. Am. Chem. Soc. **135**, 7967 (2013)
53. Chen, X.: Expanding the rule set of DNA circuitry with associative toehold activation. J. Am. Chem. Soc. **134**, 263 (2012)
54. Genot, A.J., Bath, J., Turberfield, A.J.: Combinatorial displacement of DNA strands: application to matrix multiplication and weighted sums. Angew. Chem. Int. Ed. **125**, 1227 (2013)
55. Zhu, J., Zhang, L., Dong, S., Wang, E.: Four-way junction-driven DNA strand displacement and its application in building majority logic circuit. ACS Nano **7**, 10211 (2013)
56. Picuri, J.M., Frezza, B.M., Ghadiri, M.R.: Universal translators for nucleic acid diagnosis. J. Am. Chem. Soc. **131**, 9368 (2009)
57. Tang, W., Hu, S., Wang, H., Zhao, Y., Li, N., Liu, F.: A universal molecular translator for non-nucleic acid targets that enables dynamic DNA assemblies and logic operations. Chem. Commun. **50**, 14352 (2014)
58. Zhu, J., Wang, L., Jiang, W.: A binding-induced sutured toehold activation for controllable DNA strand displacement reactions. Chem. Commun. **51**, 2903 (2015)
59. Zhu, J., Wang, L., Xu, X., Wei, H., Jiang, W.: Modular nuclease-responsive DNA Three-way junction-based dynamic assembly of a DNA device and its sensing application. Anal. Chem. **88**, 3817 (2016)
60. Li, F., Zhang, H., Lai, C., Li, X.-F., Le, X.C.: A molecular translator that acts by binding-induced DNA strand displacement for a homogeneous protein assay. Angew. Chem. Int. Ed. **51**, 9317 (2012)
61. Zhang, H., Li, F., Dever, B., Wang, C., Li, X.-F., Le, X.C.: Assembling DNA through affinity binding to achieve ultrasensitive protein detection. Angew. Chem. Int. Ed. **52**, 10698 (2013)
62. Dirks, R.M., Pierce, N.A.: Triggered amplification by hybridization chain reaction. Proc. Natl. Acad. Sci. U. S. A. **101**, 15275 (2004)
63. Bi, S., Chen, M., Jia, X., Dong, Y., Wang, Z.: Hyperbranched hybridization chain reaction for triggered signal amplification and concatenated logic circuits. Angew. Chem. Int. Ed. **54**, 8144 (2015)
64. Chandran, H., Rangnekar, A., Shetty, G., Schultes, E.A., Reif, J.H., LaBean, T.H.: An autonomously self-assembling dendritic DNA nanostructure for target DNA detection. Biotechnol. J. **8**, 221 (2013)

65. Yin, P., Choi, H.M.T., Calvert, C.R., Pierce, N.A.: Programming biomolecular self-assembly pathways. Nature **451**, 318 (2008)
66. Xuan, F., Hsing, I.M.: Triggering hairpin-free chain-branching growth of fluorescent DNA dendrimers for nonlinear hybridization chain reaction. J. Am. Chem. Soc. **136**, 9810 (2014)
67. Qian, L., Winfree, E.: A simple DNA gate motif for synthesizing large-scale circuits. J. R. Soc. Interface (2011)
68. Qian, L., Winfree, E.: Scaling up digital circuit computation with DNA strand displacement cascades. Science **332**, 1196 (2011)
69. Turberfield, A.J., Mitchell, J.C., Yurke, B., Mills, A.P., Blakey, M.I., Simmel, F.C.: DNA fuel for free-running nanomachines. Phys. Rev. Lett. **90**, 118102 (2003)
70. Muscat, R.A., Strauss, K., Ceze, L., Seelig, G.: In: Proceedings of the 40th Annual International Symposium on Computer Architecture, p. 177. ACM, Tel-Aviv, Israel (2013)
71. Dunn, K.E., Trefzer, M.A., Johnson, S., Tyrrell, A.M.: Investigating the dynamics of surface-immobilized DNA nanomachines. Sci. Rep. **6**, 29581 (2016)
72. Gu, H., Chao, J., Xiao, S.-J., Seeman, N.C.: A proximity-based programmable DNA nanoscale assembly line. Nature **465**, 202 (2010)
73. Rotaru, A., Gothelf, K.V.: DNA nanotechnology: steps towards automated synthesis. Nat. Nanotechnol. **5**, 760 (2010)
74. Wickham, S.F.J., Bath, J., Katsuda, Y., Endo, M., Hidaka, K., Sugiyama, H., Turberfield, A.J.: A DNA-based molecular motor that can navigate a network of tracks. Nat. Nanotechnol. **7**, 169 (2012)
75. Lund, K., Manzo, A.J., Dabby, N., Michelotti, N., Johnson-Buck, A., Nangreave, J., Taylor, S., Pei, R., Stojanovic, M.N., Walter, N.G., Winfree, E., Yan, H.: Molecular robots guided by prescriptive landscapes. Nature **465**, 206 (2010)
76. Chandran, H.; Gopalkrishnan, N.; Phillips, A.; Reif, J. In: Cardelli, L., Shih, W. (eds.) DNA Computing and Molecular Programming: 17th International Conference, DNA 17, Pasadena, CA, USA, September 19–23, 2011. Proceedings, p. 64. Springer Berlin Heidelberg, Berlin, Heidelberg (2011)
77. Ruiz, I.M., Arbona, J.-M., Lad, A., Mendoza, O., Aime, J.-P., Elezgaray, J.: Connecting localized DNA strand displacement reactions. Nanoscale **7**, 12970 (2015)
78. Teichmann, M., Kopperger, E., Simmel, F.C.: Robustness of localized DNA strand displacement cascades. ACS Nano **8**, 8487 (2014)
79. Peterson, A.W., Heaton, R.J., Georgiadis, R.M.: The effect of surface probe density on DNA hybridization. Nucleic Acids Res. **29**, 5163 (2001)
80. Mendoza, O., Mergny, J.-L., Aimé, J.-P., Elezgaray, J.: G-quadruplexes light up localized DNA circuits. Nano Lett. **16**, 624 (2016)
81. Dunn, K.E., Morgan, T.L., Trefzer, M.A., Johnson, S.D., Tyrrell, A.M.: In: Lones, M., Tyrrell, A., Smith, S., Fogel, G. (eds.) Information Processing in Cells and Tissues: 10th International Conference, IPCAT 2015, San Diego, CA, USA, September 14–16, 2015, Proceedings, p. 3. Springer International Publishing, Cham (2015)
82. Zhang, D.Y., Hariadi, R.F., Choi, H.M.T., Winfree, E.: Integrating DNA strand-displacement circuitry with DNA tile self-assembly. Nat. Commun. **4**, 1965 (2013)
83. Zhang, H., Li, X.-F., Le, X.C.: Binding-induced DNA assembly and its application to yoctomole detection of proteins. Anal. Chem. **84**, 877 (2012)
84. Tang, Y., Lin, Y., Yang, X., Wang, Z., Le, X.C., Li, F.: Universal strategy to engineer catalytic DNA hairpin assemblies for protein analysis. Anal. Chem. **87**, 8063 (2015)
85. Li, F., Lin, Y., Le, X.C.: Binding-induced formation of DNA three-way junctions and its application to protein detection and DNA strand displacement. Anal. Chem. **85**, 10835 (2013)
86. Li, F., Tang, Y., Traynor, S.M., Li, X.-F., Le, X.C.: Kinetics of proximity-induced intramolecular DNA strand displacement. Anal. Chem. **88**, 8152 (2016)
87. Li, F., Zhang, H., Wang, Z., Li, X., Li, X.-F., Le, X.C.: Dynamic DNA assemblies mediated by binding-induced DNA strand displacement. J. Am. Chem. Soc. **135**, 2443 (2013)
88. Tyagi, S., Kramer, F.R.: Molecular beacons: probes that fluoresce upon hybridization. Nat. Biotechnol. **14**, 303 (1996)

89. Tsourkas, A., Behlke, M.A., Rose, S.D., Bao, G.: Hybridization kinetics and thermodynamics of molecular beacons. Nucleic Acids Res. **31**, 1319 (2003)
90. Li, B., Ellington, A.D., Chen, X.: Rational, modular adaptation of enzyme-free DNA circuits to multiple detection methods. Nucleic Acids Res. **39**, e110 (2011)
91. Huang, J., Su, X., Li, Z.: Enzyme-free and amplified fluorescence DNA detection using bimolecular beacons. Anal. Chem. **84**, 5939 (2012)
92. Jiang, Z., Wang, H., Zhang, X., Liu, C., Li, Z.: An enzyme-free signal amplification strategy for sensitive detection of MicroRNA via catalyzed hairpin assembly. Anal. Methods **6**, 9477 (2014)
93. Hun, X., Xie, G., Luo, X.: Scaling up an electrochemical signal with a catalytic hairpin assembly coupling nanocatalyst label for DNA detection. Chem. Commun. **51**, 7100 (2015)
94. Chen, X., Briggs, N., McLain, J.R., Ellington, A.D.: Stacking nonenzymatic circuits for high signal gain. Proc. Natl. Acad. Sci. U. S. A. **110**, 5386 (2013)
95. Milligan, J.N., Ellington, A.D.: Using RecA protein to enhance kinetic rates of DNA circuits. Chem. Commun. **51**, 9503 (2015)
96. Li, C., Li, Y., Xu, X., Wang, X., Chen, Y., Yang, X., Liu, F., Li, N.: Fast and quantitative differentiation of single-base mismatched DNA by initial reaction rate of catalytic hairpin assembly. Biosens. Bioelectron. **60**, 57 (2014)
97. Zhang, S., Wang, K., Li, Z., Feng, Z., Sun, T.: Lab in a tube: a fast-assembled colorimetric sensor for highly sensitive detection of oligonucleotides based on a hybridization chain reaction. RSC Adv. **5**, 44714 (2015)
98. Li, B., Chen, X., Ellington, A.D.: Adapting enzyme-free DNA circuits to the detection of loop-mediated isothermal amplification reactions. Anal. Chem. **84**, 8371 (2012)
99. Jiang, Y., Li, B., Milligan, J.N., Bhadra, S., Ellington, A.D.: Real-time detection of isothermal amplification reactions with thermostable catalytic hairpin assembly. J. Am. Chem. Soc. **135**, 7430 (2013)
100. Song, W., Zhang, Q., Sun, W.: Ultrasensitive detection of nucleic acids by template enhanced hybridization followed by rolling circle amplification and catalytic hairpin assembly. Chem. Commun. **51**, 2392 (2015)
101. Dong, J., Cui, X., Deng, Y., Tang, Z.: Amplified detection of nucleic acid by G-quadruplex based hybridization chain reaction. Biosens. Bioelectron. **38**, 258 (2012)
102. Huang, J., Wu, Y., Chen, Y., Zhu, Z., Yang, X., Yang, C.J., Wang, K., Tan, W.: Pyrene-excimer probes based on the hybridization chain reaction for the detection of nucleic acids in complex biological fluids. Angew. Chem. Int. Ed. **50**, 401 (2011)
103. Liu, L., Li, Q., Tang, L.-J., Yu, R.-Q., Jiang, J.-H.: Silver nanocluster-lightened hybridization chain reaction. RSC Adv. **6**, 57502 (2016)
104. Qiu, X., Wang, P., Cao, Z.: Hybridization chain reaction modulated DNA-hosted silver nanoclusters for fluorescent identification of single nucleotide polymorphisms in the let-7 miRNA family. Biosens. Bioelectron. **60**, 351 (2014)
105. Liu, P., Yang, X., Sun, S., Wang, Q., Wang, K., Huang, J., Liu, J., He, L.: Enzyme-free colorimetric detection of DNA by using gold nanoparticles and hybridization chain reaction amplification. Anal. Chem. **85**, 7689 (2013)
106. Ma, C., Wang, W., Mulchandani, A., Shi, C.: A simple colorimetric DNA detection by target-induced hybridization chain reaction for isothermal signal amplification. Anal. Biochem. **457**, 19 (2014)
107. Chen, Y., Xu, J., Su, J., Xiang, Y., Yuan, R., Chai, Y.: In situ hybridization chain reaction amplification for universal and highly sensitive electrochemiluminescent detection of DNA. Anal. Chem. **84**, 7750 (2012)
108. Wang, X., Lau, C., Kai, M., Lu, J.: Hybridization chain reaction-based instantaneous derivatization technology for chemiluminescence detection of specific DNA sequences. Analyst **138**, 2691 (2013)
109. Niu, S., Jiang, Y., Zhang, S.: Fluorescence detection for DNA using hybridization chain reaction with enzyme-amplification. Chem. Commun. **46**, 3089 (2010)

110. Yu, X., Zhang, Z.-L., Zheng, S.-Y.: Highly sensitive DNA detection using cascade amplification strategy based on hybridization chain reaction and enzyme-induced metallization. Biosens. Bioelectron. **66**, 520 (2015)
111. Yang, D., Ning, L., Gao, T., Ye, Z., Li, G.: Enzyme-free dual amplification strategy for protein assay by coupling toehold-mediated DNA strand displacement reaction with hybridization chain reaction. Electrochem. Commun. **58**, 33 (2015)
112. Feng, C., Zhu, J., Sun, J., Jiang, W., Wang, L.: Hairpin assembly circuit-based fluorescence cooperative amplification strategy for enzyme-free and label-free detection of small molecule. Talanta **143**, 101 (2015)
113. Zou, P., Liu, Y., Wang, H., Wu, J., Zhu, F., Wu, H.: G-quadruplex DNAzyme-based chemiluminescence biosensing platform based on dual signal amplification for label-free and sensitive detection of protein. Biosens. Bioelectron. **79**, 29 (2016)
114. Chen, J., Wen, J., Yang, G., Zhou, S.: A target-induced three-way G-quadruplex Junction for 17β-Estradiol monitoring with a naked-eye readout. Chem. Commun. **51**, 12373 (2015)
115. Guo, Y., Wu, J., Ju, H.: Target-driven DNA association to initiate cyclic assembly of hairpins for biosensing and logic gate operation. Chem. Sci. **6**, 4318 (2015)
116. Kim, D., Garner, O.B., Ozcan, A., Di Carlo, D.: Homogeneous entropy-driven amplified detection of biomolecular interactions. ACS Nano **10**, 7467 (2016)
117. Zong, C., Wu, J., Liu, M., Yan, F., Ju, H.: High-throughput imaging assay of multiple proteins via target-induced DNA assembly and cleavage. Chem. Sci. **6**, 2602 (2015)
118. Choi, H.M.T., Chang, J.Y., Trinh, L.A., Padilla, J.E., Fraser, S.E., Pierce, N.A.: Programmable in situ amplification for multiplexed imaging of mRNA expression. Nat. Biotechnol. **28**, 1208 (2010)
119. Tyagi, S.: Imaging intracellular RNA distribution and dynamics in living cells. Nat. Methods **6**, 331 (2009)
120. Choi, H.M.T., Beck, V.A., Pierce, N.A.: Next-generation in situ hybridization chain reaction: higher gain, lower cost, greater durability. ACS Nano **8**, 4284 (2014)
121. Huang, J., Wang, H., Yang, X., Quan, K., Yang, Y., Ying, L., Xie, N., Ou, M., Wang, K.: Fluorescence resonance energy transfer-based hybridization chain reaction for in situ visualization of tumor-related mRNA. Chem. Sci. **7**, 3829 (2016)
122. Schweller, R.M., Zimak, J., Duose, D.Y., Qutub, A.A., Hittelman, W.N., Diehl, M.R.: Multiplexed in situ immunofluorescence using dynamic DNA complexes. Angew. Chem. Int. Ed. **51**, 9292 (2012)
123. Hemphill, J., Deiters, A.: DNA Computation in mammalian cells: microRNA logic operations. J. Am. Chem. Soc. **135**, 10512 (2013)
124. Wu, C., Cansiz, S., Zhang, L., Teng, I.T., Qiu, L., Li, J., Liu, Y., Zhou, C., Hu, R., Zhang, T., Cui, C., Cui, L., Tan, W.: A nonenzymatic hairpin DNA cascade reaction provides high signal gain of mRNA imaging inside live cells. J. Am. Chem. Soc. **137**, 4900 (2015)
125. Wu, Z., Liu, G.-Q., Yang, X.-L., Jiang, J.-H.: Electrostatic nucleic acid nanoassembly enables hybridization chain reaction in living cells for ultrasensitive mRNA imaging. J. Am. Chem. Soc. **137**, 6829 (2015)
126. Cheglakov, Z., Cronin, T.M., He, C., Weizmann, Y.: Live cell microRNA imaging using cascade hybridization reaction. J. Am. Chem. Soc. **137**, 6116 (2015)
127. Ellis, R.J.: Macromolecular crowding: an important but neglected aspect of the intracellular environment. Curr. Opin. Struc. Biol. **11**, 114 (2001)
128. Zhu, G., Zhang, S., Song, E., Zheng, J., Hu, R., Fang, X., Tan, W.: Building fluorescent DNA nanodevices on target living cell surfaces. Angew. Chem. Int. Ed. **52**, 5490 (2013)
129. Rudchenko, M., Taylor, S., Pallavi, P., Dechkovskaia, A., Khan, S., Butler Jr., V.P., Rudchenko, S., Stojanovic, M.N.: Autonomous molecular cascades for evaluation of cell surfaces. Nat. Nanotechnol. **8**, 580 (2013)
130. Han, D., Zhu, Z., Wu, C., Peng, L., Zhou, L., Gulbakan, B., Zhu, G., Williams, K.R., Tan, W.: A logical molecular circuit for programmable and autonomous regulation of protein activity using DNA aptamer-protein interactions. J. Am. Chem. Soc. **134**, 20797 (2012)

131. Han, D., Zhu, G., Wu, C., Zhu, Z., Chen, T., Zhang, X., Tan, W.: Engineering a cell-surface aptamer circuit for targeted and amplified photodynamic cancer therapy. ACS Nano **7**, 2312 (2013)
132. Wang, Y.-M., Wu, Z., Liu, S.-J., Chu, X.: Structure-switching aptamer triggering hybridization chain reaction on the cell surface for activatable theranostics. Anal. Chem. **87**, 6470 (2015)
133. Lei, Y., Tang, J., Shi, H., Ye, X., He, X., Xu, F., Yan, L.A., Qiao, Z., Wang, K.: Nature-inspired smart dna nanodoctor for activatable in vivo cancer imaging and in situ drug release based on recognition-triggered assembly of split aptamer. Anal. Chem. (2016)
134. You, M., Peng, L., Shao, N., Zhang, L., Qiu, L., Cui, C., Tan, W.: DNA "Nano-Claw": logic-based autonomous cancer targeting and therapy. J. Am. Chem. Soc. **136**, 1256 (2014)
135. You, M., Zhu, G., Chen, T., Donovan, M.J., Tan, W.: Programmable and multiparameter DNA-based logic platform for cancer recognition and targeted therapy. J. Am. Chem. Soc. **137**, 667 (2015)
136. Collins, F.S., Varmus, H.: A new initiative on precision medicine. New Engl. J. Med. **372**, 793 (2015)
137. Kusumi, A., Fujiwara, T.K., Chadda, R., Xie, M., Tsunoyama, T.A., Kalay, Z., Kasai, R.S., Suzuki, K.G.N.: Dynamic organizing principles of the plasma membrane that regulate signal transduction: commemorating the fortieth anniversary of singer and Nicolson's fluid-mosaic model. Annu. Rev. Cell Dev. Biol. **28**, 215 (2012)
138. Hooper, L.V., Littman, D.R., Macpherson, A.J.: Interactions between the microbiota and the immune system. Science **336**, 1268 (2012)
139. Wu, H.: Higher-order assemblies in a new paradigm of signal transduction. Cell **153**, 287 (2013)
140. Legrain, P., Rain, J.-C.: Twenty years of protein interaction studies for biological function deciphering. J. Proteomics **107**, 93 (2014)
141. Nam, J.-M., Stoeva, S.I., Mirkin, C.A.: Bio-bar-code-based DNA detection with PCR-like sensitivity. J. Am. Chem. Soc. **126**, 5932 (2004)
142. Nam, J.-M., Thaxton, C.S., Mirkin, C.A.: Nanoparticle-based bio-bar codes for the ultrasensitive detection of proteins. Science **301**, 1884 (2003)
143. Obernosterer, G., Martinez, J., Alenius, M.: Locked nucleic acid-based in situ detection of microRNAs in mouse tissue sections. Nat. Protoc. **2**, 1508 (2007)
144. Ngounou Wetie, A.G., Sokolowska, I., Woods, A.G., Roy, U., Loo, J.A., Darie, C.C.: Investigation of stable and transient protein-protein interactions: past, present, and future. Proteomics **13**, 538 (2013)
145. Berggård, T., Linse, S., James, P.: Methods for the detection and analysis of protein-protein interactions. Proteomics **7**, 2833 (2007)
146. Fields, S., Song, O.-K.: A novel genetic system to detect protein-protein interactions. Nature **340**, 245 (1989)
147. Morell, M., Ventura, S., Avilés, F.X.: Protein complementation assays: approaches for the in vivo analysis of protein interactions. FEBS Lett. **583**, 1684 (2009)
148. Stynen, B., Tournu, H., Tavernier, J., Van Dijck, P.: Diversity in Genetic in vivo methods for protein-protein interaction studies: from the yeast two-hybrid system to the mammalian split-luciferase system. Microbiol. Mol. Biol. R. **76**, 331 (2012)
149. Soderberg, O., Gullberg, M., Jarvius, M., Ridderstrale, K., Leuchowius, K.-J., Jarvius, J., Wester, K., Hydbring, P., Bahram, F., Larsson, L.-G., Landegren, U.: Direct observation of individual endogenous protein complexes in situ by proximity ligation. Nat. Methods **3**, 995 (2006)
150. Gullberg, M., Fredriksson, S., Taussig, M., Jarvius, J., Gustafsdottir, S., Landegren, U.: A sense of closeness: protein detection by proximity ligation. Curr. Opin. Biotechnol. **14**, 82 (2003)
151. Mocanu, M.-M., Váradi, T., Szöllősi, J., Nagy, P.: Comparative Analysis of Fluorescence Resonance Energy Transfer (FRET) and Proximity Ligation Assay (PLA). Proteomics **11**, 2063 (2011)

152. Clausson, C.-M., Arngården, L., Ishaq, O., Klaesson, A., Kühnemund, M., Grannas, K., Koos, B., Qian, X., Ranefall, P., Krzywkowski, T., Brismar, H., Nilsson, M., Wählby, C., Söderberg, O.: Compaction of rolling circle amplification products increases signal integrity and signal-to-noise ratio. Sci. Rep. **5**, 12317 (2015)
153. Clausson, C.-M., Allalou, A., Weibrecht, I., Mahmoudi, S., Farnebo, M., Landegren, U., Wahlby, C., Soderberg, O.: Increasing the dynamic range of in situ PLA. Nat. Methods **8**, 892 (2011)
154. Koos, B., Cane, G., Grannas, K., Lof, L., Arngarden, L., Heldin, J., Clausson, C.-M., Klaesson, A., Hirvonen, M.K., de Oliveira, F.M.S., Talibov, V.O., Pham, N.T., Auer, M., Danielson, U.H., Haybaeck, J., Kamali-Moghaddam, M., Soderberg, O.: Proximity-dependent initiation of hybridization chain reaction. Nat. Commun. **6**, 7294 (2015)
155. Koussa, M.A., Halvorsen, K., Ward, A., Wong, W.P.: DNA nanoswitches: a quantitative platform for gel-based biomolecular interaction analysis. Nat. Methods **12**, 123 (2015)
156. Chandrasekaran, A.R., Zavala, J., Halvorsen, K.: Programmable DNA nanoswitches for detection of nucleic acid sequences. ACS Sens. **1**, 120 (2016)
157. Hecker, K.H., Rill, R.L.: Error analysis of chemically synthesized polynucleotides. Biotechniques **24**, 256 (1998)
158. Li, B., Jiang, Y., Chen, X., Ellington, A.D.: Probing spatial organization of DNA strands using enzyme-free hairpin assembly circuits. J. Am. Chem. Soc. **134**, 13918 (2012)
159. Krueger, A., Protozanova, E., Frank-Kamenetskii, M.D.: Sequence-dependent basepair opening in DNA double helix. Biophys. J. **90**, 3091 (2006)
160. von Hippel, P.H., Johnson, N.P., Marcus, A.H.: Fifty years of DNA "Breathing": reflections on old and new approaches. Biopolymers **99**, 923 (2013)
161. Zhang, D.Y.: In: Sakakibara, Y., Mi, Y. (eds.) DNA Computing and Molecular Programming: 16th International Conference, DNA 16, Hong Kong, China, June 14–17, 2010, Revised Selected Papers, p. 162. Springer Berlin Heidelberg, Berlin, Heidelberg (2011)
162. Jiang, Y.S., Bhadra, S., Li, B., Ellington, A.D.: Mismatches improve the performance of strand-displacement nucleic acid circuits. Angew. Chem. Int. Ed. **53**, 1845 (2014)
163. Tang, Y., Wang, Z., Yang, X., Chen, J., Liu, L., Zhao, W., Le, X.C., Li, F.: Constructing real-time, wash-free, and reiterative sensors for cell surface proteins using binding-induced dynamic DNA assembly. Chem. Sci. **6**, 5729 (2015)

Chapter 3
Materials and Methods

3.1 Materials

3.1.1 DNA Oligonucleotides

All DNA oligonucleotides used in this study were purchased from Integrated DNA Technologies (IDT), and HPLC purified by IDT unless otherwise stated. All sequences used are shown in Tables A.1 - A.4. The lyophilized DNA was reconstituted in 1X Tris-EDTA buffer (1X TE, pH 8.0) to give approximately 100 μM stock and stored at 4 °C for up to a year, except for DNAs with special modifications (Cy3, Cy5 and fluorescein fluorophores, Dabcyl quencher, biotin and thiol) which were stored at −20 °C and protected from light. After reconstitution, the DNAs were quantified using a Nanodrop micro volume spectrophotometer.

3.1.2 Chemicals

The following chemicals were used as received: sodium chloride (NaCl, ≥99.5%), magnesium chloride ($MgCl_2$, ≥98%), sodium citrate tribasic dehydrate and paraformaldehyde powder (95%) were purchased from Sigma Aldrich. Potassium chloride (KCl, ≥99.5%) was purchased from AppliChem. 1X tris EDTA (TE, pH 8.0) was purchased from 1st BASE. 10X tris borate ETA (TBE, pH 8.3), 10X phosphate buffered saline (PBS, pH 7.4), and agarose powder (molecular biology grade) were purchased from Vivantis. Phosphate buffer (pH 7.4) was prepared using sodium phosphate dibasic (Na_2HPO, anhydrous enzyme grade) from Fischer Scientific and sodium dihydrogen phosphate dihydrate (H_6NaO_6P) from Kanto Chemical

A. Yan Shan, *Engineering a Robust DNA Circuit for the Direct Detection of Biomolecular Interactions*, Springer Theses,
https://doi.org/10.1007/978-981-13-2188-7_3

Co. Inc. SYBR gold nucleic acid stain (10,000X in DMSO), Hoechst 33342 Solution (20 mM) and sulfo-SMCC were purchased from Life Technologies. Tween-20 was purchased from Sinopharm. Milli-Q water (UP) with resistance >18.2 MΩ/cm was used throughout the experiment.

3.1.3 Antibodies and Biomolecules

Purified anti-human CD340 (erbB2/HER-2), purified anti-human erbB3/HER-3 and biotinylated anti-human erbB3/HER-3 were purchased from Biolegend. Biotinylated anti-ErbB2 Affibody® Molecule was purchased from Abcam. HER2/ErbB2 antibody and HER3/ErbB3 antibody used for western blot were purchased from Cell Signalling.

Albumin from bovine serum (BSA) and streptavidin (from Streptomyces Avidinii) was purchased from Sigma Aldrich. Human alpha-thrombin (8.9 mg/mL in 50% (v/v) glycerol/H_2O) was purchased from Haematologic Technologies. DNA gel loading dye (6X) and gene ruler ultra-low range DNA ladder were purchased from Life Technologies.

3.1.4 Cells and Culture Reagents

AU565 breast cancer cells (ATCC CRL-2351), MDA-MB-231 triple negative breast cancer cells (ATCC HTB-26), MCF10A normal breast epithelial cells (ATCC CRL-10317), cholera toxin and hydrocortisone were kind gifts from Prof Bay Boon Huat's lab at NUS Department of Anatomy. RPMI-1640 medium (+2.05 mM L-glutamine), DME/F12 1:1 (+2.5 mM L-glutamine), Trypsin EDTA and fetal bovine serum were purchased from GE Healthcare. Dimethyl sulfoxide (for molecular biology) and insulin solution (from bovine pancreas) were purchased from Sigma Aldrich. Recombinant human EGF (animal free) was purchased from Peprotech, USA.

3.1.5 Formulation of Buffers and Solutions

The recipe for the buffers used in this thesis are:

Standard Buffers

- 5X sodium chloride sodium citrate—tween (SSCT)
  750 mM NaCl, 75 mM trisodium citrate, 0.1% Tween-20. pH 7.0
- 1X phosphate buffered saline (PBS)
  10 mM Na_2HPO_4, 2 mM KH_2PO, 137 mM NaCl, 2.7 mM KCl. pH 7.4

- 1X tris-EDTA (TE)
  10 mM Tris-HCl, 1 mM EDTA. pH 8.0
- 1X tris borate EDTA (TBE)
  10 mM Tris-HCl, pH 8.3
- 1X Laemmli's sample buffer
  60 mM Tris-HCl (pH 6.8), 2% SDS, 10% glycerol, 5% β-mercaptoethanol, 0.01% bromophenol blue.

DNA Circuit Buffers (Self-Formulated)

- 1X thrombin binding buffer
  1X PBS, 140 mM NaCl, 5 mM $MgCl_2$, 10 mM KCl. pH 7.4
- 1X association buffer
  10 mM Tris-HCl, 480 mM NaCl, 5 mM $MgCl_2$. pH 7.0
- 1X circuit reaction buffer
  1X PBS, 10 mM $MgCl_2$. pH 7.4
- 1X blocking buffer
  1X PBS, 0.05% Tween-20, 1% (w/v) BSA. pH 7.4
- 1X washing buffer
  1X PBS, 0.05% Tween-20. pH 7.4
- 1X cell circuit buffer
  1X PBS, 0.05% Tween-20, 1% (w/v) BSA, 10 mM $MgCl_2$. pH 7.4.

3.1.6 Materials and Kits

96-well Nunc black polystyrene plate was purchased from Thermo Fisher, and 384-well black polystyrene plate with NBS™ coating and 96-well Costar black polystyrene plates with clear bottom (tissue culture treated) were purchased from Corning for analyzing biomolecules and cells respectively. For tube-based reactions, Eppendorf DNA LoBind tube (1.5 mL) was used for pure DNA reactions, Ayxgen maxymum recovery tube (1.7 mL) was used for reactions involving proteins and 0.2 mL PCR tubes from Thermo Scientific was used for annealing steps involving thermal cycler. Cell lysate extraction and subsequent protein quantification was carried out using Mem-PER plus membrane protein extraction kit and Micro BCA protein assay kit respectively from Thermo Scientific. Buffers were sterilized using Steriledflip-GP Express Plus (PES, 50 mL, 0.22 μm) from Merck. Cells were counted using In Cyto haemocytometer and cultured in 8-well Nunc chambered coverglass for confocal imaging. Hellma semi-micro quartz cuvette used for photoluminescence characterization was purchased from Sigma Aldrich.

3.2 Parameters for NUPACK Simulation

NUPACK web server was used for the design and analysis of nucleic acid structures and systems [1].

3.2.1 Designing Nucleic Acid Sequences

We defined the following input parameters in designing the nucleic acid sequences: specification of DNA or RNA desired secondary structure, domain length, sequence of domain, temperature and salt concentration. Optional parameters were also included to narrow the design space. For example, existing thrombin-binding aptamers were imposed as part of the sequence constraint and certain undesirable sequence patterns selected from a default sequence set provided by NUPACK were excluded (Table 3.1). The design code used to generate the split proximity circuit is shown in Appendix B.

3.2.2 Analysis of Nucleic Acid Systems

NUPACK analysis was carried out for *n* number of interacting DNA species (e.g. 3 for HCR analysis and 5 for split proximity circuit analysis) at 25 °C to form a maximum complex size of *n* strands. Note that this setting did not adequately represent the actual interaction in solution where higher order DNA complexes might

Table 3.1 List of standard sequence patterns which can be prevented in the nucleic acid design

Patterns	Nucleotides	Rationale	References
GGGG, AAAA, CCCC	G, A and C	To avoid self-complementary interactions, especially for promiscuous guanine	[2]
KKKKKK	Keto (G or T)	–	
MMMMMM	Amine (A or C)	–	
RRRRRR	Purine (A or G)	Long stretches polypurine and polypyrimidine could form undesired structures, e.g. triplex	[3]
YYYYYY	Pyrimidine (C or T)		
SSSSSS	Strong (C or G)	To avoid spurious hybridization and slower branch migration due to the strong base stacking effect	[4]
WWWWWW	Weak (A or T)	Weak binding leading to more breathing and lower hybridization rate constant	[4]

form, especially when HCR was involved. However, this simple analysis sufficed in capturing the initial circuit events and background noise formation. The Na^+ and Mg^{2+} concentrations were set as the same values used in the respective experiments. Key information extracted from this analysis included equilibrium concentration and Gibbs free energy of products.

3.3 Methods for DNA Circuit Preparation and Analysis

3.3.1 Preparation of DNA Components

Stock DNAs (100 μM in 1X TE, pH 8.0) were diluted to the desired working concentration for each DNA component (typically 1.0 μM for fluorescence-based experiments and 0.5*n* μM for gel electrophoresis experiments, where *n* refers to the number of circuit reaction components) in the reaction buffer for the respective systems:

- Self-contained circuit (Chap. 4): 10 mM phosphate buffer (pH 7.4), 140 mM NaCl, 10 mM KCl and 1.0 mM $MgCl_2$. KCl was added to facilitate binding of aptamers to thrombin.
- Hybridization chain reaction (Chap. 5): 5X SSCT hybridization buffer (750 mM NaCl and 75 mM trisodium citrate with 0.1% Tween-20, pH 7.0). This is a standard buffer for DNA hybridization.
- Association toehold (Chap. 6): 1X association buffer (10 mM Tris (pH 7.0), 480 mM NaCl and 5 mM $MgCl_2$.). This was optimized to produce the best signal-to-noise ratio using hairpin trigger (data not shown).
- Split proximity circuit (Chaps. 7 and 8): 1X circuit reaction buffer (1X PBS, 10 mM $MgCl_2$. pH 7.4). This buffer reflected the minimal salt concentration required for proper operation of our circuit.

All DNA strands with secondary structures, i.e. hairpins (**HP1** and **HP2**), duplex (**F-Q**) and aptamers, were heated separately to 95 °C for 5 min and left to cool to room temperature over at least 30 min (and at most 2 h) to facilitate their refolding into the most thermodynamically stable conformation. Note that the fluorescein-Dabcyl (**F-Q**) complex was prepared in a 1:2 molar ratio (on the basis of 1.0 μM **F**) to ensure complete quenching of the fluorescence. Experiments involving fluorophore-tagged DNA were always performed in the dark.

3.3.2 Reaction Procedures for Gel Electrophoresis Analysis

Equal volume of each component was mixed to obtain the final concentration of 500 nM of each circuit component and varying trigger concentrations (wherever relevant). For the negative control, e.g. 0 nM trigger concentration, the equivalent volume of reaction buffer was added instead.

The detailed sequence of addition was as follows. The trigger strand was pre-diluted in reaction buffer to the required working concentration, i.e. *n* times more concentrated than the final reaction concentration, where *n* refers to the number of circuit components. The annealed readout strands (**HP1** and **HP2**, or **F-Q**) were pre-mixed in a separate tube to their working concentrations. 2-part of the combined hairpin mixture was added to 1-part trigger solution (on volume basis) and vortexed immediately. Typically, 10 μL of each circuit components were used. The reaction was typically incubated for 1 h at room temperature (25 °C), unless otherwise stated.

3.3.3 Agarose Gel Electrophoresis Readout

2–3% agarose gel was prepared in 0.5X TBE (pH 8.3) which was pre-stained with 0.5–1X SYBR gold nucleic acid stain. SYBR gold was an appropriate stain for this project as it can stain both single stranded and double stranded DNAs which were involved in the circuit operation [5]. The formed gel slab and the 0.5X TBE running buffer was cooled down to 4 °C by incubating in fridge for 30 min. A low running temperature was preferred to reduce the effects of diffusion and obtain sharper DNA bands. The gel was run at 90 V for 40–60 min at constant voltage under ice.

3.3.4 Preparation of Antibody-DNA Conjugates Using Streptavidin Linker

Two methods were used for the preparation of antibody-DNA conjugates, one with streptavidin as the linker and the other based on the formation of covalent chemical bond using sulfo-SMCC (Appendix C). The former method of streptavidin-biotin conjugation was found to be more efficient, i.e. higher yield and no loss due to purification, and was used to produce antibody-DNA conjugates used in Chaps. 7 and 8. The bulky nature of the streptavidin-biotin conjugation did not hinder our subsequent studies.

Streptavidin was used to link the biotinylated antibody and initiators. Equivolume of 2.5 μM of biotinylated DNA and 2.5 μM of streptavidin were mixed in 1X PBS for 1 h at 37 °C before cooling down to room temperature for 30 min. Equivolume of 1.25 μM of biotinylated antibody was then added to the reaction mixture and heated to 37 °C again for 1 h before cooling down to room temperature for 30 min. The final probes were stored in 4 °C for up to 2 weeks. Unreacted streptavidin sites were blocked using 100 μM biotin in 1X PBS. It was more efficient to conjugate DNA to streptavidin first before adding antibody which resulted in approximately 100% DNA conjugation (Fig. 3.1). The reverse sequence often resulted in low DNA conjugation efficiency (approximately 10%), presumably due to steric hindrance from the larger antibody molecule (approximately 150 kDa, versus approximately 10 kDa of the DNA).

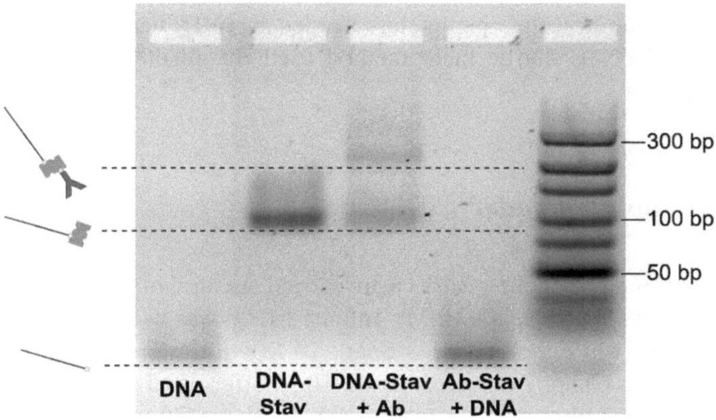

Fig. 3.1 Gel electrophoresis image of intermediate and final complexes obtained for streptavidin (Stav)-biotin conjugation in 3% agarose gel

3.4 Methods for Optical Characterization

3.4.1 UV–Vis Spectroscopy

The absorption spectrum was measured using a UV-vis spectrophotometer (Varian Cary 60). The DNA solutions were diluted until absorbance was less than 0.7 where the Beer-Lambert's law remained valid. For the calculation of quantum yield (to determine FRET efficiency in Chap. 5), the absorbance was further adjusted to less than 0.1 for wavelengths from the excitation wavelength onwards to avoid the inner filter effect [6].

3.4.2 Fluorescence Spectroscopy

The photoluminescence spectrum was measured using a fluorescence spectrophotometer (Varian Cary Eclipse). The analysis volume was kept constant at 1 mL in Hellma semi-micro quartz cuvette. Prior to any photoluminescence measurement, the fluorescent solution was always diluted such that the absorbance remained less than 0.1 and optimally within 0.02–0.08 to avoid the inner filter effect [6]. The relative quantum yield method was employed for the FRET efficiency calculation [7] using rhodamine 6G ($\Phi_F = 0.91$ in ethanol) as the reference dye [8]. The absorbance of the reference dye was adjusted to be approximately similar to that as **HP1-Cy3**. The fluorescence spectrum for the two solutions were then measured to obtain their integrated fluorescence intensities. This method was only necessary for characterizing the photoluminescence parameters needed in FRET calculation. Otherwise, microplate reader was the de facto method used for all other analysis requiring higher throughput.

3.4.3 Microplate Reader

All FRET studies and detection works were carried out on a microplate reader (Tecan Infinite M200 and Tecan Spark 10 M). Infinite M200 was used for the initial studies in this thesis, while the upgraded Spark 10 M with improved optics was used subsequently from Chap. 7 onwards. Spark 10 M offers dual optics mode, i.e. both monochromator and filter-based detection, for which the latter is more sensitive and was employed whenever appropriate filters were available.

The z-position and gain was optimized using the i-control software tool for each set of DNA readout concentration and kept constant throughout the analysis. The optimal excitation and emission wavelength pair was obtained using Tecan Infinite M200, and was kept constant for all analysis on this reader (Table 3.2). For analysis on Tecan Spark 10 M, the same set of wavelength values was used for **F-Q** readout measured in monochromator mode; while the closest set of filters was selected for HCR FRET readout measured in filter mode.

3.5 Methods for Performing Kinetics Study

The initial studies in this thesis used gel electrophoresis to evaluate the circuit kinetics which was crude and semi-quantitative at best. Following the incorporation of FRET readout for HCR in Chap. 5, circuit kinetics was tracked in real-time using fluorescence signal on microplate readers.

Table 3.2 List of excitation (λ_{Ex}) and emission (λ_{Em}) wavelengths used with the microplate readers in the respective chapters

Chapter	Reader	λ_{Ex}	λ_{Em}
5	Infinite M200	490 nm	Cy3: 568 nm Cy5: 666 nm
6	Infinite M200	478 nm	518 nm
7 and 8	Spark 10 M	FRET: 495 nm (filter-based)	Cy3: 560 nm Cy5: 665 nm
		F-Q: 478 nm (monochromator)	518 nm

3.5.1 Direct Kinetic Assay with Gel Electrophoresis Readout

This procedure was used only in Chap. 4. The kinetics for individual modules was characterized by adding their respective triggers, i.e. $c^*\ b^*$ for HCR, $t^*\ c^*$ for signal transduction and α-thrombin for target recognition, at specific time points of 0, 15, 30, 40 and 50 min to separate reaction tubes such that the respective tubes underwent reaction times of 60, 45, 30, 20 and 10 min at the end of the 1 h kinetics study. The reaction was then immediately quenched on ice and loaded onto 3% agarose gel for analysis.

3.5.2 Competitive Assay with Gel Electrophoresis Readout

This procedure was used only in Chap. 5 before FRET was incorporated as the readout signal. Quencher strands (**P1** and **P2**) were designed to block exposed domains in HCR. The main reaction solution consisted of 100 nM of **HP1** and **HP2** mixed with 10 nM of trigger strand. The reaction was quenched at respective time points by pipetting aliquots of the reaction solution into pre-prepared tubes containing two times molar ratio of **P1** and **P2** (final concentration of 200 nM). The reaction time was tightly controlled and lag time from pipetting (kept constant at 5 times up and down for each sample) was accounted for. After the last time point at 1 h, the quenched samples were loaded onto 3% agarose gel for analysis.

3.5.3 Direct Kinetic Assay with Fluorescence Readout

All analysis volume was kept constant at 100 μL in the 96-well black plate or 20 μL in the 384-well black plate. It was crucial to work with a constant analysis volume within each experimental set as the pathlength for microplate reader is effectively the height of solution in the well. The appropriate volume of reaction buffer was added to each reaction well to maintain the desired final volume. The circuit components were pre-diluted from the 1.0 μM working concentration, if needed, to 10 times the final reaction concentration. The triggers were first added to the reaction buffer. The required circuit components, i.e. **HP1**, **HP2**, **F-Q**, **I1** and/or **I2**, were pre-mixed in a separate tube before pipetting into the wells. Reaction in each well was mixed by pipetting up and down 10 times. There was lag time from well to well, and the start of reaction to actual measurement on the microplate reader. The former was accounted for by incorporating slight measurement lag time between wells. The latter was not explicitly accounted for in this thesis, though an average lag time of 2 min was often incurred. Typically, fluorescence measurement was performed every 1 min for a total analysis time of 60 min at room temperature.

3.6 Methods for Biomolecule-Based Assays

3.6.1 *One-Pot Detection of Synthetic Targets*

1.0 μM of circuit components were mixed in the following sequence to their working reaction concentration: **I1**, **HP2**, **HP1** and **I2**. The one-pot detection probes were then added directly to the targets prepared in the appropriate reaction buffer.

To study the robustness of our HCR FRET system towards biological interferences in Chap. 5, 10 μL of FBS was added to the hybridization buffer in each of the 96-well before adding the circuit components to obtain 10% FBS. The microplate reader was pre-equilibrated to 37 °C at which all reactions and measurements took place.

3.6.2 *Protein Extraction from Cells*

The Mem-PER plus membrane protein extraction kit was used to lyse cells and extract their cytosolic and membrane proteins in two separate fractions. First, approximately 5×10^6 adherent cells (AU565, MDA-MB-231 and MCF10A) were harvested in growth media by directly scraping the cells off the surface of the culture flask with a cell scraper. The harvested cells were washed to remove growth media which might interfere with the kit reagents, permeabilized to release the soluble cytosolic proteins and solubilize the membrane proteins as per manufacturer's instructions. The cytosolic and membrane protein fractions were collected separately, aliquoted and stored at −80 °C until further use. Each aliquot was only used for one freeze-thaw cycle and discarded thereafter.

3.6.3 *Analysis of Cell Lysate Protein*

Two aliquots of cell lysates were reserved for protein quantification using microBCA assay and target identification using western blot.

The microBCA assay is a modified version of the bicinchoninic acid (BCA) assay which is detergent-compatible and has been optimized for use with low protein count. The intensity of the colorimetric readout is dependent on the number of peptide bonds and the presence of four amino acids, i.e. cysteine, cystine, tryptophan and tyrosine. A series of BSA dilutions was used as the standard curve. The cell lysates often contained a high concentration of proteins and extraction reagents, and was diluted 10X and 50X with 1X PBS (pH 7.4) prior quantification. In each reaction, 10 μL of BCA reagents was added to 10 μL of sample, reacted at 60 °C in a thermocycler for 1 h and cooled down to room temperature over 10 min. The absorbance of the standards and samples was recorded at 562 nm using a Nanodrop spectrophotometer.

The standard curve (absorbance versus BSA concentration) was used to determine the protein concentration in the cell lysate samples.

Western blot experiments were conducted by Dr. Ng Cheng Teng and Dr. Chua Pei Jou from Prof Bay Boon Huat's lab at NUS Department of Anatomy.

3.6.4 One-Pot Detection of Cell Lysate

20 μL reaction volume was used in a 384-well microplate. 0.2 μg/mL of HER2-**I1**, 0.2 μg/mL of HER2-**I2** or 2.0 μg/mL of HER3-**I2,** 40 nM **HP1** and 20 nM **HP2** in 1X circuit reaction buffer were added one-pot to the cell lysate for detection of HER2:HER2 homodimers and HER2:HER3 heterodimers respectively. The original lysate was diluted by 10 times in the final reaction mixture. This was within the standard dilution range (5–50X) used in commercial kits which effectively diluted potential interference from the cell extraction buffer. Fluorescence measurement was taken at 30 min interval in a Tecan Spark 10 M reader to monitor the kinetics of signal evolution.

3.7 Methods for Cell-Based Assays

3.7.1 Cell Culture

AU565 breast cancer cells (ATCC CRL-2351) were cultured in RPMI 1640 media supplemented with 10% FBS. Culture media was changed every 3 days, while cells were passaged on a weekly basis (approximately 60% confluence) at a split ratio of 1:6–1:8 using 1X trypsin for 3 min. MDA-MB-231 triple negative breast cancer cells (ATCC HTB-26) were cultured in RPMI 1640 media supplemented with 10% FBS. Cells were passaged every three days (approximately 90% confluence) at a split ratio of 1:4 using 1X trypsin for 3 min. MCF10A normal breast epithelial cells (ATCC CRL-10317) were cultured in DMEM F12 (with L-glutamine) supplemented with 5% FBS, 20 ng/mL EGF, 100 ng/mL cholera toxin, 0.01 mg/mL insulin and 500 ng/mL hydrocortisone. Cells were passaged every three days (approximately 95% confluence) at a split ratio of 1:4 using 1X trypsin for 15 min (Note that MCF10A is highly adherent to surfaces hence the significantly longer trypsination treatment).

All cell cultures were incubated in an environment of 37 °C and 5% CO_2.

3.7.2 Cell-Based Enzyme-Linked Immunosorbent Assay (ELISA)

ELISA is a well-established immunochemistry method used to detect and quantify specific antigen or antibody in a sample. In this project, cells were seeded at a density of approximately 1.5×10^5 cells/cm^2 in a 96-well black microplate with clear bottom overnight to achieve monolayer coverage. They were then washed with sterile 1X PBS and fixed with 4% paraformaldehyde for 15 min to crosslink within the cell structure and to the microplate. After which, the cells were washed thoroughly 3 times using sterile 1X PBS, each time with 5 min incubation to remove unreacted paraformaldehyde completely. The wells were then replaced with 1X blocking buffer and left to incubate for 2 h with shaking.

For the standard ELISA protocol used to establish the binding affinity of the antibodies and confirm the proper operation of the DNA circuit, the primary antibodies recognizing HER2 and/or HER3 were added to the cells at varying concentrations (0.2–2.0 μg/mL) and incubated at 4 °C overnight in 1X blocking buffer with shaking. The wells were washed 3 times with 1X washing buffer. For immunofluorescence experiments where the antibodies were tagged with streptavidin-Cy3, the microplate was used directly for fluorescence measurement. Otherwise, a mixture of 40 nM **HP1** and 20 nM **HP2** in 1X circuit reaction buffer was then added to develop HCR FRET signals.

For one-pot detection using our circuit, the microplate was washed once with 1X washing buffer after the blocking step. This was followed by the one-pot addition of 0.2 μg/mL of HER2-**I1**, 0.2 μg/mL of HER2-**I2** or 2.0 μg/mL of HER3-**I2**, 40 nM **HP1** and 20 nM **HP2** in 1X cell reaction buffer. HCR signals were measured directly without further washing.

3.7.3 Cell Imaging on Confocal Microscopy

The cells were seeded at a density of 0.5×10^5 cells/cm^2 in an 8-well chambered coverglass (Nunc™ Lab-Tek™) overnight. A lower seeding density was used so that individual cells could be imaged. The cells were washed and fixed with 4% paraformaldehyde for 15 min. After washing with 1X PBS, the chamber was incubated with 1X blocking buffer for 2 h with shaking to reduce non-specific binding.

While blocking the cell chamber, the circuit probes were prepared as follows: 1 μM of **HP1** and **HP2** were annealed in 1X cell circuit buffer; while the active streptavidin sites on HER2-**I1**, HER2-**I2** and HER3-**I2** were blocked with 100 μM free biotin in 1X PBS. Just before adding to the cells, the circuit components were mixed in the following sequence in 1X cell circuit buffer: HER2-**I1**, **HP2**, **HP1** and HER2-**I2** or HER3-**I2**.

After 2 h of blocking, the blocking buffer was removed and the chamber was washed once with 1X PBS. The cells were then incubated with 1 μg/mL of Hoechst dye (diluted with 1X PBS from a working concentration of 1 mg/mL immediately before use) for 5 min, followed by washing three times with 1X PBS. The prepared DNA circuit mixture was then added to the cells in one-pot and imaged directly under a Nikon A1 confocal microscope for kinetics study. Otherwise, the reaction was allowed for proceed for 30 min before pipetting the circuit mixture from the chamber and replacing with 1X cell circuit buffer. This effectively quenched the circuit reaction at the 30 min mark which we found to be sufficient for signal development from the kinetics study and to ensure comparable reaction time across all chambers.

The Nikon A1 laser scanning confocal system was used for FRET, time-lapse and z-stack imaging. To achieve FRET, the donor (Cy3) was excited using 488 nm Argon gas laser and the acceptor (Cy5) emission was collected using Cy5 filter wheel. Cell nucleus stained with Hoechst dye was excited with 405 nm diode laser and the emission collected with DAPI filter wheel. Most images were captured using 40X objective to survey a more representative cell population. Close-up images at 100X oil objective were taken to confirm structural details. The gain and laser power was kept constant for all confocal imaging. The captured images were analyzed using ImageJ with the same brightness and contrast adjustment [9].

3.8 Data Analysis and Interpretation

3.8.1 Analyzing Gel Electrophoresis Readout

The signal intensity of the hybridization chain reaction (HCR) products and the unpolymerized hairpin (**HP1** and **HP2**) bands inputted was quantified using ImageJ [9].

A parameter of relative signal intensity (rSI) was coined in Chap. 4 to determine the percentage of polymerized hairpins and was calculated by the ratio of HCR product signal intensity to the total signal intensity in each lane (taken to be total amount of input hairpins). Signal-to-background ratio (S/B) was calculated by dividing the rSI obtained in presence of α-thrombin (positive signal) to that in absence of α-thrombin (background noise).

To characterize the HCR kinetics in Chap. 5, only the signal intensity of the hairpin monomer band (I_t) was used for analysis. This was because the intensity of DNA-binding dyes tends to increase non-linearly with the DNA molecular weight. While this inherent signal amplification was good for detection, it complicated the analysis of HCR kinetics. The extent of reaction (ξ) was calculated from the hairpin intensity at $t = 0$ (I_o) (Eq. 3.1).

$$\xi = \frac{I_0 - I_t}{I_0} \tag{3.1}$$

We simplified HCR as a bimolecular reaction between the trigger and hairpin (taken as a single entity). This did not reflect the much more complex hybridization kinetics involved in the cascaded opening of hairpin monomers in HCR, but sufficed as a simple model for us to quantitatively compare the kinetics of different hairpin designs. The extent of reaction was fitted as a pseudo-first order reaction since the concentration of trigger was lower than that of the hairpins.

3.8.2 Analyzing Fluorophore-Quencher (F-Q) System

In each experiment, a negative control was established using the same circuit components as in the test samples but without the trigger/target, which we termed as the background noise. A baseline correction (F_o) was established using a blank where the same volume of pure reaction buffer was used. A positive control was established using 100 nM of **F-Q** and 150 nM of direct trigger (**DT**) to completely displace all quencher (**Q**) strands. The fluorescence intensity obtained here was taken to correspond to 100 nM of released fluorophores (F_{max}).

All raw fluorescence readings (F) were corrected for baseline fluctuations ($F - F_0$), followed by normalizing against F_{max} to calculate the relative fluorescence signals (RFU), i.e. 1 RFU was equivalent to 1 nM of released fluorophore. This minimized the random fluctuation of signals due to, e.g. instrument noise and environment. The signal-to-background (S/B) ratio was defined as ($RFU_{signal} - RFU_{background}$)/$RFU_{background}$, where the signal and background were generated in presence and absence of the trigger/target respectively.

Graphs used for the detailed characterization of minute leakages were shown without error bars (highlighted in their respective legends) to avoid congesting the narrow RFU range represented. Instead, we used scatter plots to represent the mean value for triplicate experiments, which were then fitted with a solid linear trend line to guide the eyes in identifying time-dependent trends.

3.8.3 Analyzing HCR FRET Systems

The FRET ratio was determined by taking the ratio of Cy5 (F_{Cy5}) to Cy3 (F_{Cy3}) emission for a given excitation wavelength. The background emission was subtracted at the respective wavelengths by exciting a blank well containing the same volume of reaction buffer without any DNA species, $F_{dye,0}$. The FRET ratio is defined in Eq. 3.2.

$$FRET\ Ratio = \frac{F_{Cy5} - F_{Cy5,0}}{F_{Cy3} - F_{Cy3,0}} \quad (3.2)$$

Limit of detection (LOD) was calculated based on the signal level three times S.D. above the mean of negative control. Negative control referred to the reaction with all circuit components except the target. The molar equivalent of LOD was calculated from the linear regression between the signal and dosage.

3.8.4 Calculation of Reaction Kinetics Parameters

We modelled the circuit operation as a second order reaction as typical for DNA hybridization though this just provides a simplified picture of the actual reaction involving multiple intermediate steps.

The reaction kinetics studies were typically designed as pseudo-first order reaction by keeping the trigger strand as the limiting reactant, i.e. one-tenth of the circuit components. The kinetics of signals generated (be it the intensity of gel band, **F-Q** fluorescence intensity or FRET ratio) was fitted to a first order reaction to obtain the pseudo first order rate constant (k'). The actual second-order rate constant (k) was calculated by multiplying k' by the concentration of the excess reactant.

References

1. Zadeh, J.N., Steenberg, C.D., Bois, J.S., Wolfe, B.R., Pierce, M.B., Khan, A.R., Dirks, R.M., Pierce, N.A.: NUPACK: analysis and design of nucleic acid systems. J. Comput. Chem. **32**, 170 (2011)
2. SantaLucia, J., Hicks, D.: The thermodynamics of DNA structural motifs. Annu. Rev. Biophys. Biomol. Struct. **33**, 415 (2004)
3. Michel, D., Chatelain, G., Herault, Y., Brun, G.: The long repetitive polypurine/polypyrimidine sequence (TTCCC)48 Forms DNA triplex with PU-PU-PY base triplets in vivo. Nucleic Acids Res. **20**, 439 (1992)
4. Zhang, D.Y., Winfree, E.: Control of DNA strand displacement kinetics using toehold exchange. J. Am. Chem. Soc. **131**, 17303 (2009)
5. Tuma, R.S., Beaudet, M.P., Jin, X., Jones, L.J., Cheung, C.-Y., Yue, S., Singer, V.L.: Characterization of Sybr gold nucleic acid gel stain: a dye optimized for use with 300-nm Ultraviolet transilluminators. Anal. Biochem. **268**, 278 (1999)
6. Lakowicz, J.R.: Principles of Fluorescence Spectroscopy. Springer, US (2007)
7. Williams, A.T.R., Winfield, S.A., Miller, J.N.: Relative fluorescence quantum yields using a computer-controlled luminescence spectrometer. Analyst **108**, 1067 (1983)
8. Würth, C., Grabolle, M., Pauli, J., Spieles, M., Resch-Genger, U.: Relative and absolute determination of fluorescence quantum yields of transparent samples. Nat. Protocols **8**, 1535 (2013)
9. Schneider, C.A., Rasband, W.S., Eliceiri, K.W.: NIH image to ImageJ: 25 years of image analysis. Nat. Methods **9**, 671 (2012)

Chapter 4
Modular Framework for Engineering a Self-contained DNA Circuit

4.1 Introduction

This chapter sets up the modular framework from which each module will be further developed in subsequent chapters. The focus of this chapter is to demonstrate the autonomous and modular nature of our designed circuit, while highlighting key design concepts and criteria underlying the final embodiment of the split proximity circuit.

We are interested in the system of biomolecular interactions where two or more entities come into close physical proximity to elicit certain biological effect(s). We conceptualized an "AND" logic gate which takes in three inputs of the first and second binding event, and that they must occur within close proximity, for which the range of separation distance is a user-defined value, to perform this computation (Scheme 4.1). The signal output should remain localized at the binding site to detect the interaction event(s) in situ.

The desired circuit operation was then translated into functional structures and domains from which computation was performed via a series of toehold-mediated strand displacement. α-Thrombin was selected as the model protein as it has two well-characterized DNA aptamer sequences known to bind to the fibrinogen- and heparin-recognition exosites located at opposite ends of the molecule [1].

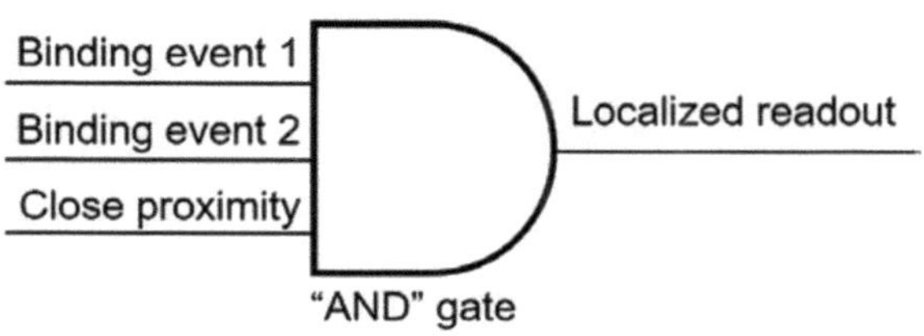

Scheme 4.1 The simplest form of the proposed circuit for detecting biomolecular interactions can be expressed as a "AND" gate

A. Yan Shan, *Engineering a Robust DNA Circuit for the Direct Detection of Biomolecular Interactions*, Springer Theses,
https://doi.org/10.1007/978-981-13-2188-7_4

We designed two initiator strands for the two binding recognition events. Each initiator contained a "key" region which was designed to be locked up at equilibrium so that no signal developed under static condition (Scheme 4.2). The first initiator (**I1**) was an aptamer-based molecular switch which could undergo conformational change upon thrombin binding to expose the first "key". This unlocked the second "key" on the second initiator (**I2**) bound to the other exosite by hybridizing with the protector (**P**) strand when **I1** and **I2-P** were in proximity with each other. The hybridization chain reaction (HCR) (Sect. 2.3.3) was then triggered to form long DNA chain on the α-thrombin molecule.

Every step in the dynamic assembly process involved direct physical contact among the circuit components which remained bound to the target binding sites where all circuit operations took place. Having the circuit in such a self-contained format signal cross-talk by free-moving DNA strands and facilitated the development of localized signal. An important advantage of the close proximity criteria is the drastic increase of the local DNA concentration which in turn enhanced the strand exchange kinetics [2]. The free DNA concentration in solution is typically ranged from nanomolar to at most micromolar. By confining the circuit operation within the immediate space surrounding α-thrombin molecule, which can be approximated as a sphere of 5 nm diameter [3], the DNA concentration was effectively reduced to one probe per control volume of radius approximately 20 nm (considering the size of α-thrombin and the length of fully-stretched initiators), or approximately 50 mM. This represents a 5–7 order of magnitude enhancement of DNA concentration as compared to the free species in solution.

Finally, the circuit domains were translated into DNA sequences using NUPACK simulation [4]. While NUPACK is a simple and powerful tool for designing duplexes and hairpins, it does not account for non-Watson Crick base pairing, such as the G-quadruplex structure of thrombin-binding aptamers [5]. As such, we had to first test different molecular switch designs through experimental trial-and-error. Once the

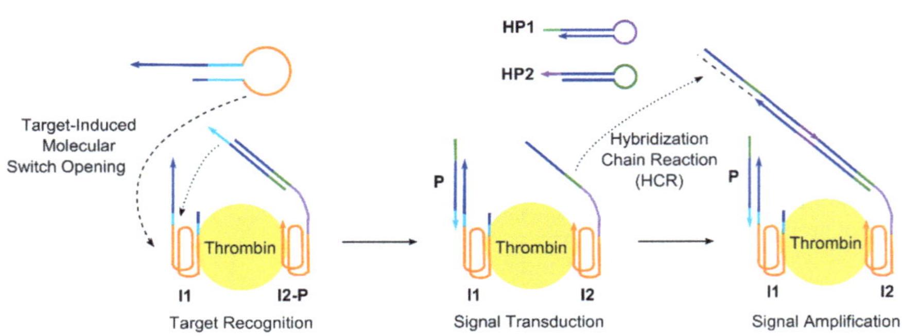

Scheme 4.2 Schematic outline of the dynamic assembly process involved in the proposed self-contained DNA circuit for detecting proximity-based biomolecular interactions. All DNA components were designed to be bound onto the α-thrombin molecule during their participation in the autonomous assembly process

circuit was fully defined, the kinetics of individual modules were characterized, and the reaction condition was further optimized based on the rate-limiting step which in this case was α-thrombin binding. The complete circuit operation was then evaluated, followed by a specific demonstration of the advantage of having an increased local probe concentration in a self-contained circuit.

4.2 Results and Discussion

4.2.1 Detailed Breakdown of the DNA Circuit Domain Design

The circuit components presented in Scheme 4.2 were further broken down into the domain structure in Scheme 4.3 and the actual sequences are found in Appendix A. The key (domain *t**) of initiator 1 (**I1**) was initially locked up in the stem of a hairpin molecular switch which opened in response to the specific aptamer recognition of α-thrombin by domain *a*1. **I2** was bound to α-thrombin via domain *a*2. The key (domain *b**) of initiator 2 (**I2**), originally locked up by a partially complementary protector (**P**) strand, was then unlocked by the released key of **I1**. Following signal transduction, a second key (domain *b**) was exposed to trigger HCR involving the cascaded opening of two hairpin monomers (**HP1** and **HP2**) via alternating toehold domains *b* and *d* [6]. The length of the spacer (domain *s*) in **I2** could be varied to probe separation distances up to approximately 30 nm. A "clamp" (domain *e*) was included in **I1** to combat spurious opening of the molecular switch [7]. This overall circuit was designed to operate by capturing proximity events and translating them into a localized signal comprising of long DNA chain.

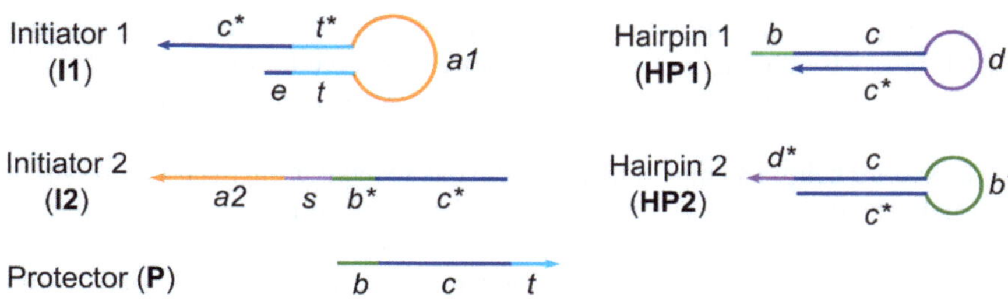

Scheme 4.3 The detailed breakdown of the domains involved in the respective circuit components is shown. There are four main circuit components made up of 5 DNA strands. The two "keys" (domains *t** and *b**) were carefully locked up to effect autonomous strand displacements leading to HCR readout in response to dual binding events taking place on a single α-thrombin molecule

4.2.2 Design of Target Binding-Induced Molecular Switch

As the two triggering processes, i.e. α-thrombin binding-induced hairpin opening and initiation of HCR, were independent of each other, we could design the initiator sequences separately. The original HCR hairpin sequences reported by Pierce's group were used, hence fixing domains *b*, *c* and *d*. This leaves the molecular switch design for **I1** as the only circuit component to be optimized. Specifically, this refers to the design of domain *t* ("key") and *e* ("clamp"), and the relative positioning of the domains within a molecular switch structure.

Though several hairpin molecular switch designs for α-thrombin binding aptamers were previously reported, [8–10] the additional presence of toehold in the circuit introduced competing hybridization by its complementary domain which might favour untriggered hairpin opening. For instances, domain *t* in **P** could bind to domain t^* in **I1** when the end sequence "breathed" [11]. It was necessary to optimize the molecular switch design based on the following two criteria: 1) being able to lock up the "key" in **I1** sufficiently against hybridization pressure from the complementary domain to minimize background leakage while 2) being sufficiently metastable to open and bind to α-thrombin. The drastic conformation change from a duplex structure held in a hairpin loop to the thrombin-binding G-quadruplex structure then revealed the sequestered toehold domain.

Five molecular switch designs were considered, each varying in terms of the toehold length, clamping sequence length and position of the aptamer sequence. The basic idea was to lock part of the aptamer sequence as the "key" domain in the hairpin stem. The remaining aptamer sequence was left in the loop region to initiate molecular reconfiguration from α-thrombin binding and pull the hairpin stem apart. The first consideration was the length of the hairpin stem (which was also the "key")—a longer length favored both hairpin stability and the competing "unlocking" strand displacement by domain *t*. By comparing Design i (Lanes 6 and 7) and Design ii (Lanes 10 and 11) in Fig. 4.1a, the more prominent effect was on hairpin stabilization as HCR products of higher molecular weight was observed for the 9 nt length which reflected lesser amount of triggering. Still, it was not sufficient to suppress circuit leakages.

The shorter hairpin stem was more effective in reconfiguring itself in response to α-thrombin binding and so, we retained the 5 nt length while adding a "clamp" domain in Design iii and iv to improve hairpin stability without promoting the "unlocking" process. We also considered placing the aptamer sequence at the end of the hairpin stem in Design v (Lane 16 and 17) to avoid the kinetically slow process of hairpin opening which turned out to be an ineffective strategy.

The only design that remained locked in absence of α-thrombin yet triggered HCR in presence of α-thrombin was Design iii (Lane 12 and 13). The "clamp" domain was the crucial element in buffering the transient exposure of the "key" domain when the ends of the stem "breathed" as in Design i and Design ii. The clamp length greatly influenced the outcome of this strategy as excessive length led to overly stable hairpin as observed for Design iv (Lanes 14 and 15). This hindered the hairpin opening

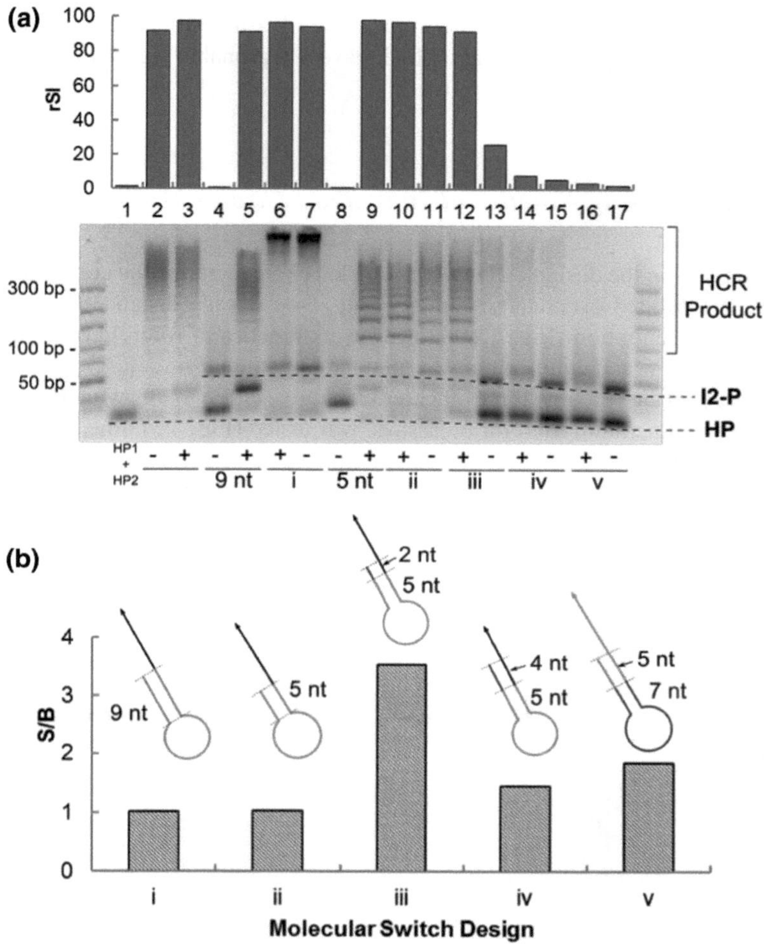

Fig. 4.1 Screening for the best molecular switch design amongst five design candidates. **a** Gel electrophoresis image of the performance of the five molecular switch candidates (denoted as "i" to "v"), with "-" and "+" denoting the absence and presence of α-thrombin respectively. Lanes 2 and 3 correspond to the case where initiator 1 was replaced by a strand designed to be fully complementary to the protector strand. Lanes 4 and 5 correspond to the negative and positive control for toehold length of 9 nt while lanes 8 and 9 correspond to that for toehold length of 5 nt. A 10–300 bp DNA ladder is shown at the left side of the gel. The relative signal intensity (rSI) of the HCR products formed is indicated above the respective lanes. **b** The signal-to-background ratio (S/B) of the HCR products formed in presence to that in absence of α-thrombin was calculated for each molecular switch design (depicted above each corresponding graph). Design iii gave the best S/B ratio of approximately 3.5

process such that no HCR occurred in presence of α-thrombin. Thus, we selected Design iii for the rest of this study which gave a reasonable signal-to-background ratio of approximately 3.5 (Fig. 4.1b).

4.2.3 Evaluating Kinetics of Individual Circuit Modules

After confirming the design of **I1**, we proceeded to characterize the complete circuit design. It exhibited excellent specificity and some dosage dependence on the α-thrombin concentration (Fig. 4.2). However, its kinetics at room temperature was exceedingly slow and equilibrium was attained only after 7 h.

To speed up the reaction, three other temperatures within 37 °C, which is typically the highest temperature used for bioassays, was tested (Fig. 4.3). Ideally the circuit should generate meaningful results on the time scale of minutes, and we selected 30 min and 1 h as two evaluation time points. As the temperature increased from 30 to 37 °C, more HCR products formed within 30 min relative to the 1 h reference time point. Notably, the background leakage was not significant within the analysis time of 1 h even at the highest temperature (37 °C) used. Thus, 37 °C was used for subsequent studies.

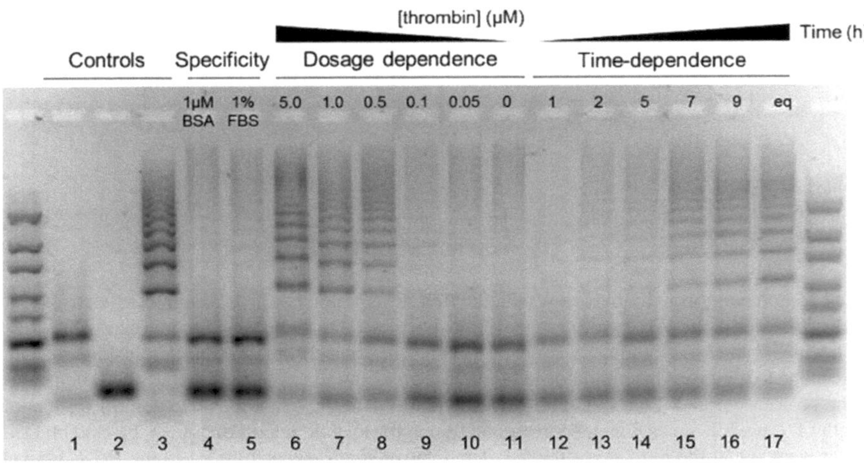

Fig. 4.2 Gel electrophoresis image of the performance of the self-contained DNA circuit evaluated at RT. Lane 1 corresponds to hairpins (**HP1** and **HP2**) only; lane 2 corresponds to hairpins and initiator 2—protector complex; lane 3 corresponds to that in lane 2 with an additional trigger strand. Hybridization chain reaction (HCR) was not triggered by non-specific protein (1 µM BSA, Lane 4) and interfering matrix (10% FBS, Lane 5). Lanes 6–11 represent the signal developed at different α-thrombin concentrations as labelled. Lanes 12–17 represent the kinetics of signal development where the equilibrium amount of HCR products was attained only after approximately 7 h

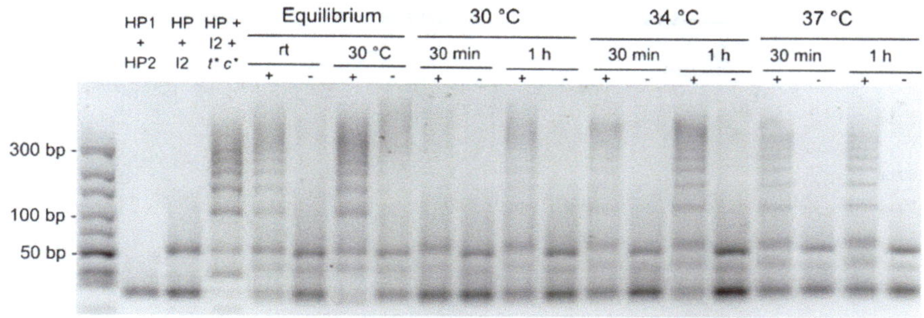

Fig. 4.3 Gel electrophoresis image demonstrating the use of temperature to modulate the circuit kinetics. The reaction was analyzed after 30 min and 1 h for three temperatures studied (30, 34 and 37 °C). "+" and "-" denotes the presence and absence of α-thrombin respectively. A 10–300 bp DNA ladder is shown at the left side of the gel

After establishing the reaction temperature, the circuit was deconstructed into individual steps involved in the key modules: (i) signal amplification—HCR, (ii) signal transduction—toehold-mediated strand displacement, and (iii) α-thrombin recognition—molecular switch opening, to evaluate their respective kinetics. The minimum DNA components required for each module to operate and the necessary add-ons when progressing from one module to the next during the re-construction process are shown in Scheme 4.4.

For (i) HCR, $c^*\,b^*$ strand sufficed in triggering the cascaded opening of hairpin 1 (**HP1**) and hairpin 2 (**HP2**). To progress to (ii) signal transduction, a protector (**P**) strand was added to the $c^*\,b^*$ strand to block HCR which was subsequently triggered only in presence of an additional $t^*\,c^*$ strand. To ensure that the signal developed only in the presence of (iii) specific target recognition, two aptamer sequences (domains $a1$ and $a2$) were added to the $t^*\,c^*$ and $c^*\,b^*$ strands to obtain the complete initiator 1 (**I1**) and initiator 2 (**I2**) probes. The binding of α-thrombin induced the opening of **I1** to initiate signal transduction followed by HCR in the final re-constructed circuit.

The individual deconstructed modules were tested at selected time points to semi-quantitatively evaluate their respective kinetics (Fig. 4.4a). Both HCR and the addi-

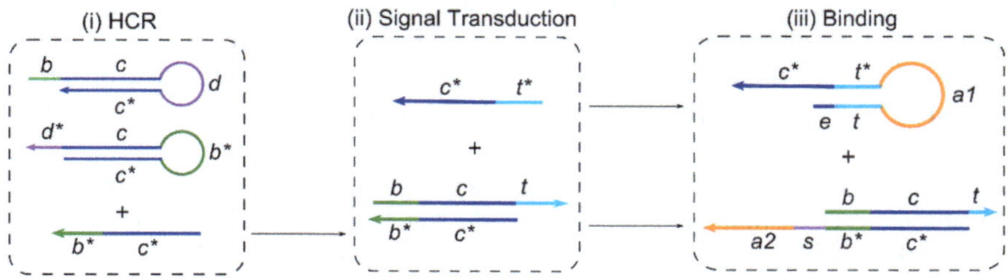

Scheme 4.4 Each module was decomposed into the minimum DNA components required for each step to operate

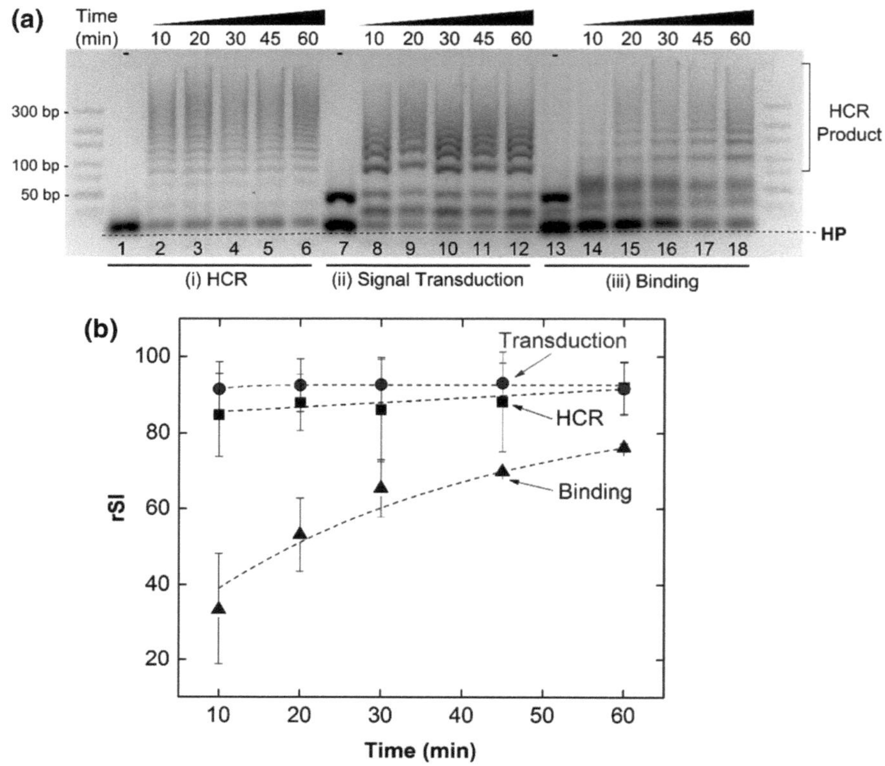

Fig. 4.4 The molecular switch opening step was found to be rate limiting. **a** Gel electrophoresis image to characterize the kinetics of individual module as indicated below the gel. The reaction proceeded for the time period specified at 37 °C. Lanes 1, 7 and 13 correspond to the negative control for each module where no trigger was added, i.e. c^* b^* strand for HCR, t^* c^* strand for signal transduction and α-thrombin for target binding. **b** The additional α-thrombin binding step slowed the circuit kinetics considerably, compared to the other two steps of signal transduction and HCR. All data are shown as mean ± standard deviation (n = 3). rSI = relative signal intensity

tional signal transduction step exhibited fast kinetics and equilibrium signal was attained by 10 min (Fig. 4.4b). However, having an additional binding step slowed down the circuit operation considerably. This suggests that the reaction was limited by the opening of the molecular switch. Nonetheless, the overall circuit operation time was greatly reduced from approximately 7 h at room temperature to the current approximately 30 min.

It is worth noting that the buffer condition chosen for the circuit operation was a physiologically-relevant condition of 140 mM NaCl, 1 mM $MgCl_2$ and 10 mM KCl in 10 mM phosphate buffer (pH 7.4) to ensure that the circuit remained compatible with other biomolecules or cell systems. Despite using a considerably lower salt condition than those typically used for pure DNA circuitry studies (Sect. 2.5.3) which could have disadvantaged our reaction kinetics and thermodynamics, the relatively short analysis time of 30 min and approximately 70% of HCR signal developed at equilibrium was encouraging.

4.2.4 *Evaluating Overall Circuit Performance*

After characterizing the individual DNA circuit components, the overall circuit performance in terms of detection limit and selectivity was formally evaluated (Fig. 4.5a). In this proof-of-concept work, gel electrophoresis was used as the main readout tool to visualize and semi-quantify the amount of HCR products formed. The one-pot circuit and α-thrombin mixture was loaded directly into the agarose gel without further manipulation steps. The lack of distinct HCR products in presence of non-specific protein (1 μM bovine serum albumin (BSA), Lane 6) and interference matrix (1% fetal bovine serum (FBS), Lane 7, and 10% FBS, Lane 8) demonstrated the good selectivity of the DNA circuit. It should be noted that the selectivity is closely linked to the intrinsic performance of the aptamer sequence and may have varied performance for different targets. HCR products could form in the presence of a mixture of α-thrombin and the same set of interferences mentioned above (Lane 9–11), which demonstrated the robustness of the DNA circuit.

The HCR signal exhibited dosage dependence on the changing concentration of α-thrombin within a physiologically relevant range of 50 nM—5 μM (Lane 12–16) [12]. Semi-quantitative analysis was carried out by measuring the relative signal intensity (rSI) of the HCR products in each gel lane. From Fig. 4.5b, higher signal intensity developed with increasing α-thrombin concentration which started to plateau off at 1 μM and above.

Since a linear dosage-signal relationship was obtained for α-thrombin concentration of 50 nM—1 μM, the proximity-based circuit could be used for quantification (Fig. 4.5b). The limit of detection (LOD) for this assay was found to be 50 nM in its current form. A lower LOD is achievable using fluorescence signal as the readout which will be covered in the next chapter. We will also show in Chap. 8 how the fluorescence signal can be used for in situ visualization at the interaction sites.

4.2.5 *Feasibility of Enhancing the Development of Localized Signal Through Dual Binding Events*

Since the goal of the proposed circuit is for the direct detection of biomolecular interaction, it is imperative that localized signal forms only in response to multiple bio-recognition events. However, it was possible for the unbound **I2-P** complex to undergo signal transduction with the bound **I1** strand and form freely moving HCR products (Scheme 4.5), which is commonly encountered in the design of local-

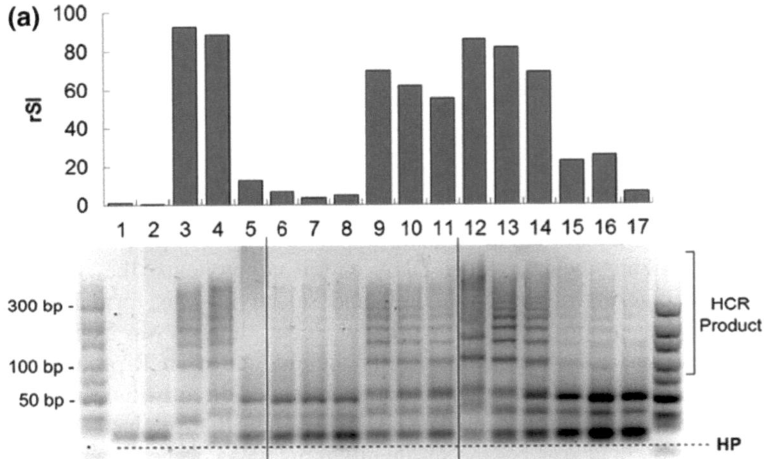

	HP1 + HP2	HP + I2-P	HP + I2-P + I*c*	Eq	Eq	Selectivity						Sensitivity					
[Thrombin] (μM)	0	0	0	1.0	0	0	0	0	1.0	1.0	1.0	5.0	1.0	0.5	0.1	0.05	0
Others	-					1 μM BSA	1% FBS	10% FBS	1 μM BSA	1% FBS	10% FBS	-					

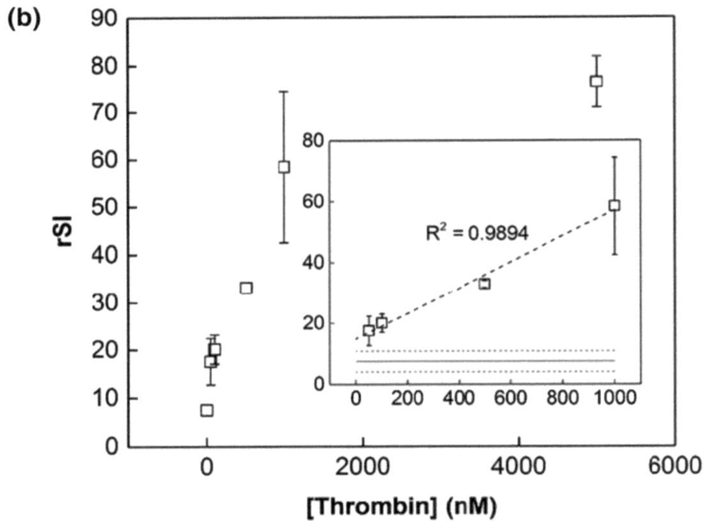

Fig. 4.5 The performance of the self-contained circuit was evaluated. **a** Gel electrophoresis image of the performance of the circuit in terms of both selectivity and sensitivity evaluated at 37 °C. A 10–300 bp DNA ladder is shown at both sides of the gel. The relative signal intensity (rSI) of the HCR products formed is indicated above the respective lanes. **b** The HCR signal developed exhibited dosage-dependence for a physiologically-relevant α-thrombin concentration ranging from 50 nM to 5 μM. Inset: A linear dosage-signal relationship ($R^2 = 0.9894$) was obtained for 50 nM to 1 μM of α-thrombin. The solid line represents the mean background noise in absence of α-thrombin while the dashed lines represent ± 3 σ of this mean. All data are shown as mean ± standard deviation (n = 3)

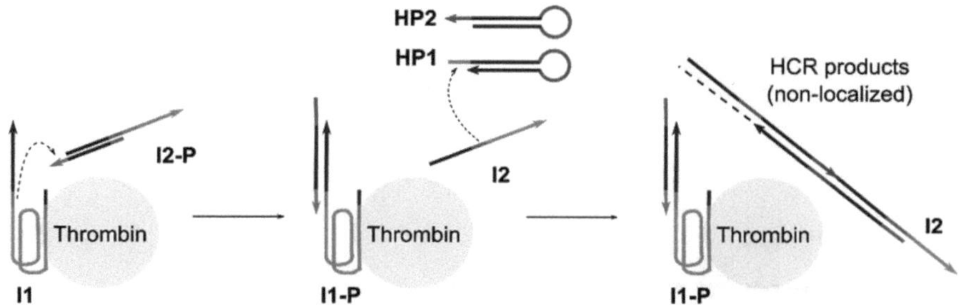

Scheme 4.5 Formation of non-localized HCR products when α-thrombin-bound **I1** hybridizes with unbound **I2-P** for signal transduction

ized circuits [13]. For the purpose of detecting and visualizing interaction activities directly on top of just end-point detection, this possibility posed two challenges to the circuit designed—first, the occurrence of false-positives and next, the development of non-localized signal.

Here, we re-visited the concept of increased local concentration due to the co-binding of two initiator strands on a single α-thrombin molecule, which should promote a faster rate of signal transduction as compared to the free moving, unbound **I2-P**. A control **I2** strand without the aptamer sequence (**noApt**) was designed to prevent binding to α-thrombin. This represented the undesired circuit operation taking place between α-thrombin-bound **I1** and unbound **I2**.

The first strategy was to kinetically subdue the undesired operation involving unbound **I2** by cutting down on two kinetics-related parameters, i.e. **I2** concentration and analysis time. The next strategy was to further enhance the rate of local signal transduction upon dual initiators binding on a single α-thrombin molecule. One way was to reduce the spacer length on **I2** and hence the effective reaction volume. One can picture on a molecular level where a shorter spacer spatially constrains the degree of free movement of **I2**, while **I2** with a longer spacer moves more flexibly and takes a longer time to approach **I1**. To test this concept, two spacer lengths were used in **I2**–15 nt (**I2-15T**; this was the design used for all previous studies) and 5 nt (**I2-5T**).

As the concentration of **I2-P** concentration increased, more HCR products formed (Fig. 4.6). In all cases, the bound probes (**I2-15T** and **I2-5T**) triggered more HCR signal than the unbound **noApt** probe. The undesired, non-localized HCR products generated from the unbound probes was observed more prominently at higher **I2-P** concentration, e.g. more undesired HCR products was seen in Lane 19 than Lane 7. A good balance between maximizing signal production and minimizing noise generation was achieved at roughly 0.3:1 **I2-P**:**I1** ratio and this initiator ratio was used for the next part of the study.

The circuit operation was next analyzed at time points of 10, 20, 30, 45 and 60 min. For analysis shorter than 30 min, distinct HCR products formed for bound **I2** (first 2 lanes for each time point) while the signal for **noApt** (3rd lane for each time point) remained relatively weak (Fig. 4.7a). Since the non-localized reaction

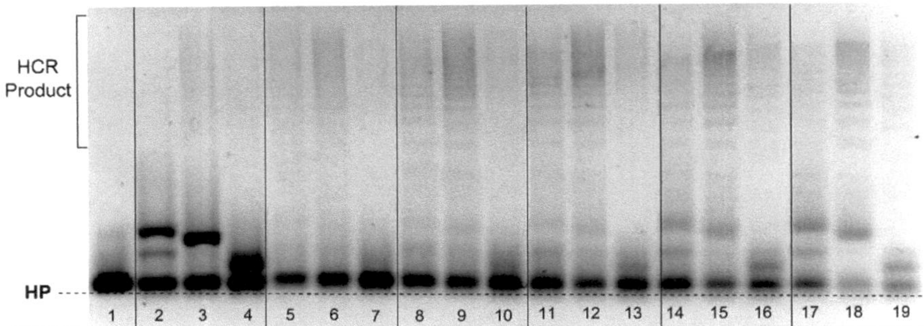

	1	2	3	4	5	6	7	8	9	10	11	12	13	14	15	16	17	18	19
I2-P : I1	-	1 : 1			0.1 : 1			0.2 : 1			0.3 : 1			0.5 : 1			1 : 1		
I2	-	I2-15T	I2-5T	noApt	I2-15T	I2-5T	noApt	I2-15T	I2-5T	noApt	I2-15T	I2-5T	noApt	I2-15T	I2-5T	noApt	I2-15T	I2-5T	noApt
[Thrombin]	HP1 + HP2	0 μM			1.0 μM														

Fig. 4.6 Gel electrophoresis image to show the effect of initiator 2–P (**I2-P**) concentration on the relative localized-to-non-localized signal ratio. Different ratios of **I2-P** to **I1** were used while keeping **I1** concentration constant as 1.0 μM for all cases. The analysis time was kept constant at 15 min for all concentration ratios. Lane 1 corresponds to hairpins (**HP1** and **HP2**) only; while lanes 2–4 correspond to the background signal in absence of α-thrombin

appears to grow linearly with time for all three designs, it was best to keep the reaction time within 20–30 min, where the generation of localized signal by **I2-5T** was still statistically higher ($p < 0.05$) than the non-localized signal by **noApt** (Fig. 4.7b).

The highest absolute signal was consistently obtained for **I2-5T**, over **I2-15T** and **noApt**, presumably due to the faster initial reaction rate (unable to quantify in this study using gel electrophoresis). This observation supported our idea that greater spatial confinement was beneficial for the formation of localized signal. Hence, the shortest possible spacer length should be used to enhance the selectivity of localized over non-localized reaction, so long as **I1** and **I2** were still able to physically approach each other for that separation distance. The maximum separation distance that our circuit is theoretically capable of detecting is approximately 30 nm by increasing the spacer length to approximately 100 nt (near the limit of conventional solid-phase oligonucleotide synthesis technology); however, the enhancement effect of local concentration will diminish accordingly.

We envision that our DNA circuit design can be applied for cell surface sensing. The idea of cell typing via DNA computation of cell-surface receptors was conceptualized by Stojanovic's [14] and Tan's [15] groups separately (Sect. 2.4.3), where a one-target-to-one-signal design was adopted. The localized signal amplification scheme featured in this chapter can potentially enhance the signal from individual targets for the direct visualization of cell surface markers with varying abundance. These could be disease biomarkers existing either as single target, e.g. CCR5, a co-receptor in HIV-1 infection, [16] or as higher order oligomers, e.g. HER2:HER2 homodimers and HER2:HER3 heterodimers on breast cancer cells [17]. We will demonstrate the latter application in Chap. 8.

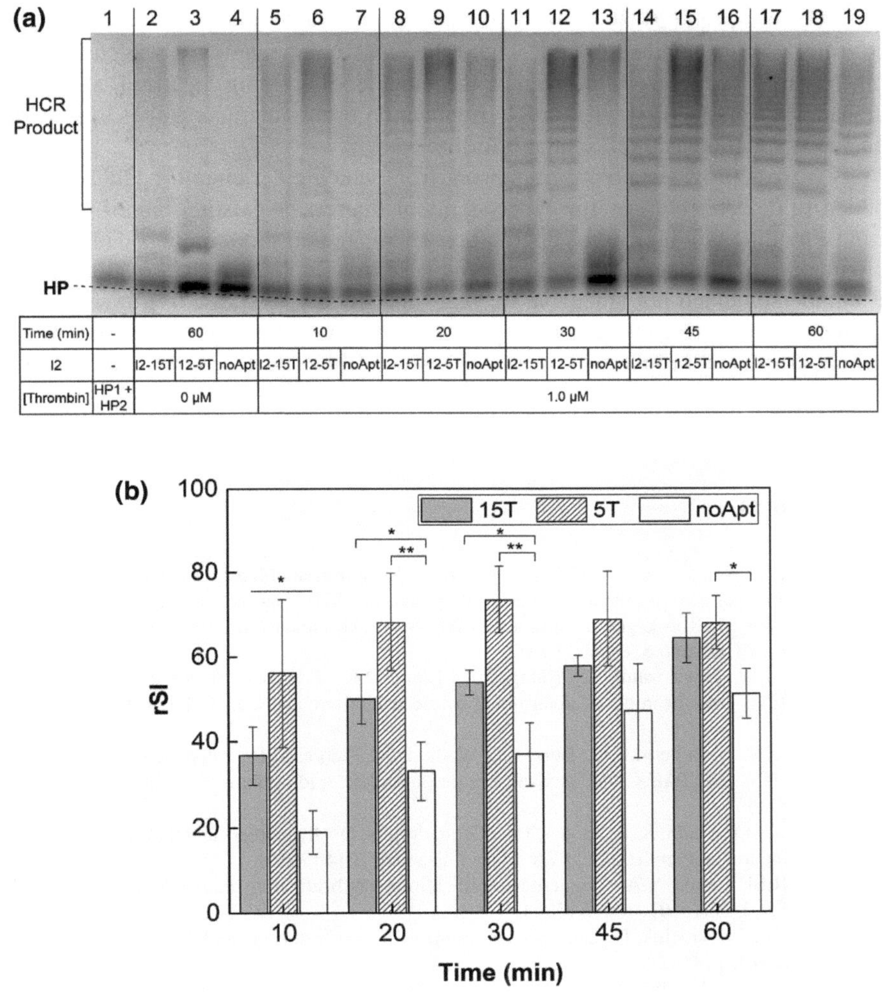

Fig. 4.7 The spacer length was used to control the development of localized over non-localized signals. **a** Gel electrophoresis image of the HCR products developed at the indicated time points of analysis. **b** The readout signal increased with time for all 3 designs. The effect of analysis time and the use of bound/unbound **I2** was found to be statistically significant using two-factor ANOVA test ($\alpha = 0.05$, n = 3). Pair-wise comparisons between the unbound (**noApt**) and bound (**I2-15T** or **I2-5T**) cases at each time points were performed using one-tailed Student's t-test (* $= p < 0.1$, ** $= p < 0.05$.). All data are shown as mean ± standard error (n = 3). rSI = relative signal intensity

4.3 Conclusion

In conclusion, a self-contained DNA circuit capable of sensing proximity-based interaction events was designed and demonstrated using α-thrombin as a model protein. The circuit performed with good selectivity, reasonable limit of detection

(50 nM) and speed (20–30 min). More importantly, we demonstrated the advantage of using a self-contained circuit, where the DNA components assembled directly on the target recognized, to enhance the kinetics of forming localized, amplified signal. This can facilitate the direct visualization of the interaction sites to provide an additional dimension of spatial information.

This work provides a basic framework from which other circuit modules can be added to detect other specific bio-systems of interest or existing circuit modules can be modified to improve its operability. There are two shortcomings with current circuit design—first, gel electrophoresis was used to visualize the HCR readout which is cumbersome and not amenable for real-time studies; next, the use of molecular switch as a triggering event is an elegant strategy but is limited to targets with well-validated aptamer sequences. These two issues will be addressed in Chaps. 5 and 6 respectively.

References

1. Rinker, S., Ke, Y., Liu, Y., Chhabra, R., Yan, H.: Self-assembled DNA nanostructures for distance-dependent multivalent ligand-protein binding. Nat. Nanotechonol. **3**, 418 (2008)
2. Teichmann, M., Kopperger, E., Simmel, F.C.: Robustness of localized DNA strand displacement cascades. ACS Nano **8**, 8487 (2014)
3. Weisel, J.W., Nagaswami, C., Young, T.A., Light, D.R.: The shape of thrombomodulin and interactions with thrombin as determined by electron microscopy. J. Biol. Chem. **271**, 31485 (1996)
4. Zadeh, J.N., Steenberg, C.D., Bois, J.S., Wolfe, B.R., Pierce, M.B., Khan, A.R., Dirks, R.M., Pierce, N.A.: NUPACK: analysis and design of nucleic acid systems. J. Comput. Chem. **32**, 170 (2011)
5. Tucker, W.O., Shum, K.T., Tanner, J.A.: G-quadruplex DNA aptamers and their ligands: structure, function and application. Curr. Pharm. Des. **18**, 2014 (2012)
6. Dirks, R.M., Pierce, N.A.: Triggered amplification by hybridization chain reaction. Proc. Natl. Acad. Sci. U.S.A. **101**, 15275 (2004)
7. Seelig, G., Soloveichik, D., Zhang, D.Y., Winfree, E.: Enzyme-free nucleic acid logic circuits. Science **314**, 1585 (2006)
8. Li, J.J., Fang, X., Tan, W.: Molecular aptamer beacons for real-time protein recognition. Biochem. Biophys. Res. Commun. **292**, 31 (2002)
9. Bang, G.S., Cho, S., Kim, B.-G.: A novel electrochemical detection method for aptamer biosensors. Biosens. Bioelectron. **21**, 863 (2005)
10. Xue, L., Zhou, X., Xing, D.: Sensitive and homogeneous protein detection based on target-triggered aptamer hairpin switch and nicking enzyme assisted fluorescence signal amplification. Anal. Chem. **84**, 3507 (2012)
11. Li, B., Ellington, A.D., Chen, X.: Rational, modular adaptation of enzyme-free DNA circuits to multiple detection methods. Nucleic Acids Res. **39**, e110 (2011)
12. Lee, M., Walt, D.R.: A fiber-optic microarray biosensor using aptamers as receptors. Anal. Biochem. **282**, 142 (2000)
13. Mendoza, O., Mergny, J.-L., Aimé, J.-P., Elezgaray, J.: G-quadruplexes light up localized DNA circuits. Nano Lett. **16**, 624 (2016)
14. Rudchenko, M., Taylor, S., Pallavi, P., Dechkovskaia, A., Khan, S., Butler Jr., V.P., Rudchenko, S., Stojanovic, M.N.: Autonomous molecular cascades for evaluation of cell surfaces. Nat. Nanotechnol. **8**, 580 (2013)

15. You, M., Peng, L., Shao, N., Zhang, L., Qiu, L., Cui, C., Tan, W.: DNA "nano-claw": logic-based autonomous cancer targeting and therapy. J. Am. Chem. Soc. **136**, 1256 (2014)
16. Simon, V., Ho, D.D.: Hiv-1 dynamics in vivo: implications for therapy. Nat. Rev. Micro. **1**, 181 (2003)
17. de Paula Costa Monteiro, I., Madureira, P., de Vasconscelos, A., Humberto Pozza, D., Andrade de Mello, R.: Targeting HER family In HER2-positive metastatic breast cancer: potential biomarkers and novel targeted therapies. Pharmacogenomics **16**, 257 (2015)

Chapter 5
Designing Hybridization Chain Reaction Monomers for Robust Signal Amplification

5.1 Introduction

We have shown that hybridization chain reaction (HCR) can be used in our self-contained circuit as a robust, selective and sensitive signal amplification tool in Chap. 4. There were two issues which we aim to address and improve in this chapter. The first aim is to re-design robust HCR hairpins with minimal leakages so that the circuit design is not constrained by pre-existing HCR domains. The next aim is to incorporate fluorescence signal as HCR readout for real-time detection while preserving the one-pot and self-contained features of our circuit.

The original hairpin sequences reported by Piece's group were used in the previous chapter. Though they have performed well across many studies [1–3], the hardwiring of specific hairpin domains reduced the degree of freedom in the overall circuit design. We believe that one reason why such a simple, enzyme-free amplification tool has yet to attain widespread adoption is the lack of explicit design guidelines to instruct the rational re-design of hairpins. The design process is a non-trivial task due to challenge posed by circuit leakages [4] which was made more serious by the amplification mechanism of HCR. While useful strategies to suppress background leakage have been proposed for another amplification scheme—catalytic hairpin assembly (CHA) [5], they are not directly applicable to HCR due to the different arrangement of domains. To the best of our knowledge, the best way to achieve high HCR signal gain is to start with a good hairpin sequence design [6].

After designing a set of working hairpins, an appropriate readout signal needs to be incorporated. Essentially, HCR results in the formation of a longer DNA chain from short hairpin monomers. It is challenging to devise a simple readout strategy based on the change in molecular weight. Gel electrophoresis is one method which we used in the previous chapter; however, it is not practical for actual detection works nor useful for real-time studies. Alternative strategies often involve additional separation steps to remove unhybridized hairpins which lengthen the operation time and increase the likelihood of error arising from additional manipulation steps [7–11]. Other research

A. Yan Shan, *Engineering a Robust DNA Circuit for the Direct Detection of Biomolecular Interactions*, Springer Theses,
https://doi.org/10.1007/978-981-13-2188-7_5

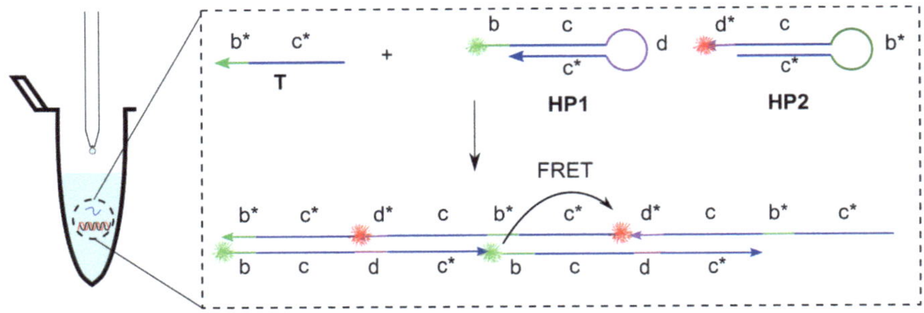

Scheme 5.1 Schematics of the one-pot HCR scheme where FRET readout was incorporated

groups have worked around the constraint through creative molecular designs, e.g. via incorporation of G-quadruplex sequences [12] and using gold nanoparticle colorimetric readout [12, 13]. While still achieving the intended amplification effect, these strategies have foregone an important feature of HCR as a powerful visualization tool in producing localized signals.

We approached these two issues by first establishing a framework of general design guidelines for the lengths and G·C content of the hairpin toehold and stem domains. Trends were obtained from NUPACK simulations and certain sequences were selected to validate these trends experimentally. The design guidelines were then implemented to design hairpins with notably shorter stem length, i.e. two-third of Pierce's design. Having a shorter hairpin stem length resulted in faster HCR kinetics, while our guidelines ensured that the hairpins remained sufficiently robust to counter the increased risk of circuit leakage.

Next, since the separation distance between the hybridized hairpins was reduced, a distance-sensitive Förster resonance energy transfer (FRET) readout signal was incorporated by tagging individual hairpins with a pair of fluorophores (Scheme 5.1). This readout format did not require any separation step and retained the one-pot feature of our circuit. Also, the signal remained tethered to the target(s) for localized amplification. The ratiometric nature of FRET measurement minimized background interferences.

5.2 Results and Discussion

5.2.1 NUPACK Simulation of Hairpin Toehold and Stem Requirements

Four main factors were considered for designing the hairpin sequences, i.e. length and G·C content of both toehold and stem domains. Pierce's original hairpin sequence design used a stem length of 18 nucleotides (nt) to lock a 6 nt active toehold [14], which is significantly longer than the 4–6 nt typically used to stabilize a hairpin molecular switch [15]. Even though DNA circuits are known to suffer from the issue of circuit leakage [5, 16], we questioned if such excessive stem length were necessary.

Trends were identified by screening hairpin sequences with various domain configurations through NUPACK simulation [17]. Each of the four factors was studied as a variable in the simulation to elucidate their limiting boundaries. The essence of the simulation work is summarized in Appendix D. We will focus the main discussion around Pierce's designs in this chapter itself to communicate the key points more clearly. The original HCR hairpin sequences reported [14] was evolved gradually with the key objective of minimizing background leakages. Here, we define background leakage as any complexes formed from **HP1** and/or **HP2** in absence of **DT**.

First, the excessively long stem length (18 nt) could be drastically reduced to similar length as the toehold (6 nt) without appreciable background leakage according to NUPACK simulation (Fig. 5.1). The metastability of the hairpins was less sensitive to stem length than the stem G·C content especially when the stem length approached that of the toehold. A rough cut-off of more than 60% stem G·C content was effective in suppressing the background leakage.

The hairpin metastability was more sensitive to the properties of toehold than stem domain. Generally, the G·C content in the toehold should be kept below approximately 30–40% regardless of the toehold length, which must be shorter than the stem (Table 5.1). 12 nt was taken as the upper limit of the toehold length as recommended by Pierce's group to prevent stable complexes from forming between the toehold of one hairpin and the loop of the other hairpin [6].

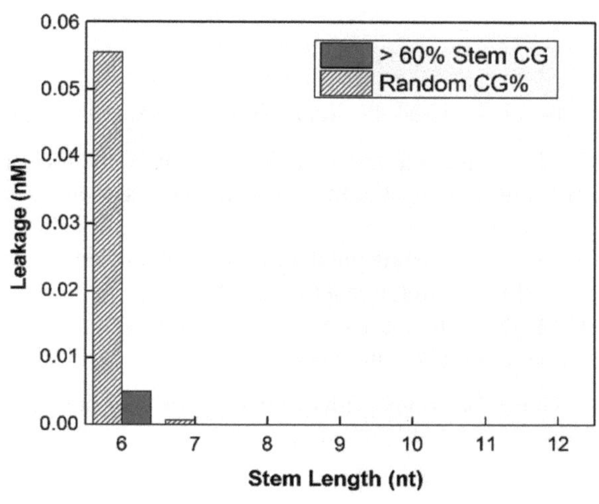

Fig. 5.1 Effect of stem length and G·C content on background leakage. Stem length was less crucial than G·C content in maintaining the hairpin metastability. The toehold length was kept constant as 6 nt

Table 5.1 Effect of toehold length and C·G content (%) on the amount of background leakage (nM). The stem length was kept constant as 9 nt with a C·G content of 60%. The NUPACK simulation was based on **HP1** and **HP2** concentration of 500 nM, and $[Na^{+}]=0.75$ nM and $[Mg^{2+}]=0$

Length	C·G %	DNA Sequence	Leak (nM)
6	33.3	**HP1** TTAACCGCCGAATCCCAAAGTGGATTCGGC **HP2** GGATTCGGCGGTTAAGCCGAATCCACTTTG **T** GGATTCGGCGGTTAA	0
	50.0	**HP1** CTAACCGCCGAATCCCAAAGTGGATTCGGC **HP2** GGATTCGGCGGTTAGGCCGAATCCACTTTG **T** GGATTCGGCGGTTAG	0
7	28.6	**HP1** TTTAACCGCCGAATCCCAAAGTGGATTCGGC **HP2** GGATTCGGCGGTTAAAGCCGAATCCACTTTG **T** GGATTCGGCGGTTAAA	0
	42.8	**HP1** CTTAACCGCCGAATCCCAAAGTGGATTCGGC **HP2** GGATTCGGCGGTTAAGGCCGAATCCACTTTG **T** GGATTCGGCGGTTAAG	2.4
8	25.0	**HP1** TTTTAACCGCCGAATCCCAAAGTGGATTCGGC **HP2** GGATTCGGCGGTTAAAAGCCGAATCCACTTTG **T** GGATTCGGCGGTTAAAA	0
	50.0	**HP1** CCTTAACCGCCGAATCCCAAAGTGGATTCGGC **HP2** GGATTCGGCGGTTAAGGGCCGAATCCACTTTG **T** GGATTCGGCGGTTAAGG	0.55
9	22.2	**HP1** TTTTTAACCGCCGAATCCCAAAGTGGATTCGGC **HP2** GGATTCGGCGGTTAAAAAGCCGAATCCACTTTG **T** GGATTCGGCGGTTAAAAA	0
	33.3	**HP1** CTTTTAACCGCCGAATCCCAAAGTGGATTCGGC **HP2** GGATTCGGCGGTTAAAAGGCCGAATCCACTTTG **T** GGATTCGGCGGTTAAAAG	0.7
10	10.0	**HP1** TTTTTTAATCGCCGAATCCCAAAGTGGATTCGGC **HP2** GGATTCGGCGATTAAAAAAGCCGAATCCACTTTG **T** GGATTCGGCGATTAAAAA	0.0
	20.0	**HP1** TTTTTTAACCGCCGAATCCCAAAGTGGATTCGGC **HP2** GGATTCGGCGGTTAAAAAAGCCGAATCCACTTTG **T** GGATTCGGCGGTTAAAAA	1.6

We crystallized the simulation results into the four-point design guidelines:

(1) It is critical to keep the G·C content in the toehold domain below 30–40%.
(2) Hairpin metastability is sensitive to the toehold length which should not exceed 12 nt.
(3) Any stem length greater than the toehold length can maintain the hairpin metastability if Guideline #4 is fulfilled.
(4) If the stem and toehold lengths are comparable, a minimum of 60% stem G·C content is recommended.

These four points fully define the critical hairpin structures, i.e. length and G·C content of toehold and stem domains.

5.2.2 Experimental Validation of Simulation Results

To test the design guidelines experimentally, we first reduced the stem length of Pierce's original design by a third to 12 nt. No background leakage was observed (Fig. 5.2). Next, a new set of hairpin sequences (**HP1** and **HP2**) was generated randomly using NUPACK based on the same hairpin structure and G·C content. The hairpin candidates were then screened manually based on the guidelines listed in Sect. 5.2.1. No background leakage was observed when all the guidelines were adhered to as observed from the negative controls (0 nM trigger lanes) in Fig. 5.2.

These design guidelines are not meant to be all-encompassing and serve as a rational starting point for the flexible design of HCR hairpins of various toehold and stem lengths, from which minor refinements may still be required for specific sequences. At the point of writing this thesis, 15 sets of HCR hairpin sequences had been validated experimentally for their robustness against leakages and responsiveness towards triggering (Table 5.2). The validity of all HCR hairpin sequences was evaluated using gel electrophoresis which involved high DNA concentrations; an adverse condition that promotes circuit leakage.

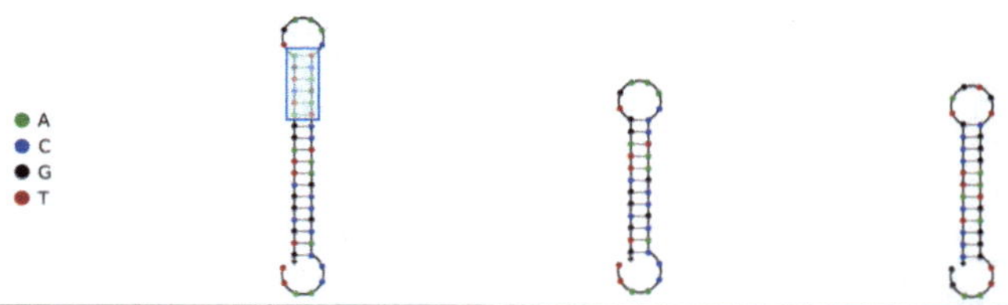

	Pierce (Original)	Pierce (Modified)	New Sequence
Stem length	18 nt	12 nt	12 nt
Stem CG%	55%	66.7%	66.7%

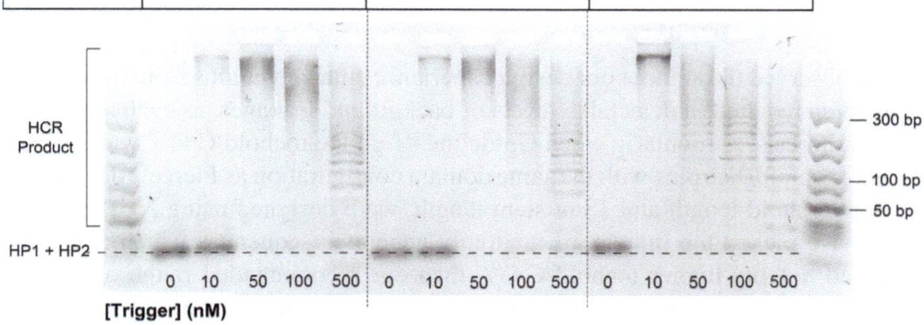

Fig. 5.2 Evolution of HCR hairpin design by reducing the stem length. Gel electrophoresis analysis showed improved HCR performance when a shorter stem length was used without generating discernible background noise. The hybridization reaction was performed in increasing trigger concentration (0, 10, 50, 100, 500 nM) at 25 °C. Specifications (stem length and C·G %) of the representative HCR hairpin designs were shown above the respective lanes. A 10–300 bp DNA ladder is shown on both sides of the gel

Table 5.2 List of HCR hairpin 1 (**HP1**) and hairpin 2 (**HP2**) sequences designed using NUPACK based on our guidelines and validated experimentally

Hairpin	Sequences
HP1	1: GTT GGA ATT GGG AGT AAG GGC TGT GAT GCC CTT ACT CCC
	2: CCT TCA TTA CCC TCA CAC TCC TGC CAT GGA GTG TGA GGG
	3: CAC CTT AAA CCT CAC CCT ACC TCA AGT GGT AGG GTG AGG
	4: GGA ATT GGG AGT AAG GGC TGT GAT GCC CTT ACT CCC
	5: TGG AAT AAC GGG AGT AAG GGC TGT GAT GCC CTT ACT CCC
	6: TGG AAT TGG GAG TAA GGG CTG TGA TGC CCT TAC TCC C
	7: TTG GAA TTG GGA GTA AGG GCT GTG ATG CCC TTA CTC CC
	8: GAG TTT AGA GCG AGT GAC CTG TGC ATT CAG GTC ACT CGC
	9: TTA ACC CAC GCC GAA TCC CAA AGT GGA TTC GGC GTG
	10: GTT GGA ATT GGG AGT AAG GGC ATG TGA TGC CCT TAC TCC C
	11: GTT GGA ATT GGG AGT AAG GGC AAT GTG ATG CCC TTA CTC CC
	12: GTT GGA ATT GGG AGT AAG GGC GAA TGT GAT GCC CTT ACT CCC
	13: GTT GGA ATT GGG AGT AGG CTG TGA TGC CTA CTC CC
	14: TTA ACC CCG ATC CCA AAG TGG ATC GG
	15: TTA ACC GCC GAA TCC CAA AGT GGA TTC GGC
HP2	1: GCC CTT ACT CCC AAT TCC AAC GGG AGT AAG GGC ATC ACA
	2: GGA GTG TGA GGG TAA TGA AGG CCC TCA CAC TCC ATG GCA
	3: GGT AGG GTG AGG TTT AAG GTG CCT CAC CCT ACC ACT TGA
	4: GCC CTT ACT CCC AAT TCC GGG AGT AAG GGC ATC ACA
	5: GCC CTT ACT CCC GTT ATT CCA GGG AGT AAG GGC ATC ACA
	6: GCC CTT ACT CCC AAT TCC AGG GAG TAA GGG CAT CAC A
	7: GCC CTT ACT CCC AAT TCC AAG GGA GTA AGG GCA TCA CA
	8: CAG GTC ACT CGC TCT AAA CTC GCG AGT GAC CTG AAT GCA
	9: GGA TTC GGC GTG GGT TAA CAC GCC GAA TCC ACT TTG
	10: GCC CTT ACT CCC AAT TCC AAC GGG AGT AAG GGC ATC ACA T
	11: GCC CTT ACT CCC AAT TCC AAC GGG AGT AAG GGC ATC ACA TT
	12: GCC CTT ACT CCC AAT TCC AAC GGG AGT AAG GGC ATC ACA TTC
	13: GCC TAC TCC CAA TTC CAA CGG GAG TAG GCA TCA CA
	14: GGA TCG GGG TTA ACC GAT CCA CTT TG
	15: GGA TTC GGC GGT TAA GCC GAA TCC ACT TTG

We observed throughout our design experiences that deviations from these guidelines often resulted in detectable levels of background leakages, as evaluated on gel electrophoresis, particularly when Guideline #1 on the toehold G·C% was violated. In one example, hairpins with the same domain configuration as Pierce's design [14], i.e. 6 nt toehold length and 18 nt stem length, were designed using NUPACK web server. We picked the first set of randomly-generated sequences which coincidentally had notably higher toehold C·G% than our recommended range of 30–40% (Fig. 5.3). Significant circuit leakage was observed in Lane 6 where no trigger was added. Readers may notice that the stem C·G% was also lower than our recommended >60% cut-off. It is worth pointing out at this point that Pierce's original design had the same stem C·G% of 55.6%. This was consistent with our experimental observations to date that the stem C·G% was a less crucial factor when the hairpin stem was significantly longer than the toehold length by a heuristically derived cut-off of two times.

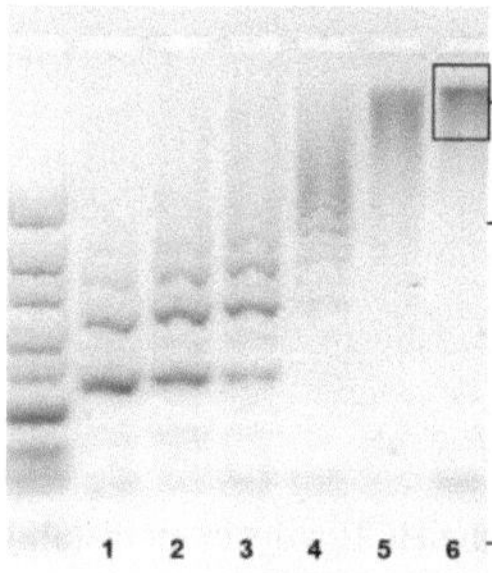

Strand	Sequence	Toehold CG%	Stem CG%
HP1	GAT GAG GGT AGT GAG GGT GAG ATG GGC TTG CAT CTC ACC CTC ACT ACC	50.0	55.6
HP2	CAT CTC ACC CTC ACT ACC CTC ATC GGT AGT GAG GGT GAG ATG CAA GCC	66.7	55.6
Trigger	CAT CTC ACC CTC ACT ACC CTC ATC	-	-

Fig. 5.3 Gel electrophoresis image for hairpins of 18 nt stem and 6 nt toehold with C·G content outside the recommended range. Lanes 1–6 represent decreasing trigger concentration of 1000, 500, 100, 50, 10 and 0 nM (negative control). 1.0 μM of **HP1** and **HP2** were used. The background leakage is indicated by the black box. A 10–300 bp DNA ladder is shown on the left-hand side of the gel. The DNA sequences and corresponding toehold/stem C·G% are shown in the right-hand side table

It is encouraging to note that all design attempts based on the guidelines stated in Sect. 5.2.1 had been successful thus far, and we are looking forward to validating a larger pool of sequences. It is also worth noting that the original sequences [14] and four out of the five improved sequences [6] reported by Pierce's group are in line with the above guidelines. We believe these guidelines will be useful to others who are interested in using HCR for specific downstream applications without having to divert time towards understanding the black box of hairpin design [18].

5.2.3 Robustness of Design Guidelines Under Various Experimental Conditions

The robustness of our newly-generated HCR hairpin sequences was challenged under different experimental conditions of salt concentrations and reaction temperatures. Though only one set of sequences was tested explicitly in this section, we wish to qualify that most of the other hairpin sequences listed in Table 5.2 did not experience leakages when used under different environments.

The only exception was the #14 hairpin set which had an exceptionally short stem length of 7 nt and toehold length of 6 nt. In this case, noticeable leakage developed over longer analysis time of more than 6 h under high salt condition of 5× SSCT and high hairpin concentrations of 500 nM. Nonetheless, its metastability was preserved under lower salt condition in an overnight analysis. This represents a limiting case and it is unlikely for one to design a hairpin of comparable stem length to the toehold which will suffer from lower duplex stability in its hybridized state and therefore penalize HCR progress.

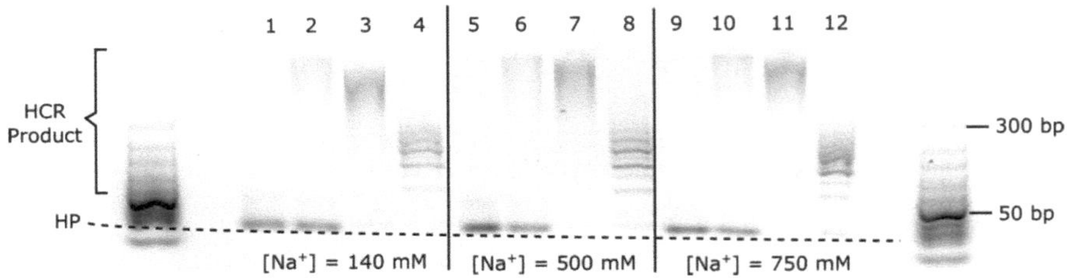

Fig. 5.4 Gel image demonstrating the effect of NaCl concentration on the HCR hairpin metastability. Different buffers of varying NaCl concentration commonly encountered in HCR studies were used: 1× PBS (pH 7.4) (Lanes 1–4), 50 mM H_2PO_4 (pH 6.8) and 500 mM NaCl (Lanes 5–8) and 5× SSCT (Lanes 9–12). For each buffer system, trigger concentration of (from left to right) 0×, 0.01×, 0.1× and 1.0× that of 500 nM hairpins was used. The reaction mixture was incubated at 25 °C for 1 h. A 10–300 bp DNA ladder is shown on both sides of the gel

5.2.3.1 Sodium Chloride (NaCl) Concentration

Three common buffer systems for HCR studies containing 140 mM [19], 500 mM [14] and 750 mM [6] NaCl respectively was used evaluate the effect of NaCl concentration. From Fig. 5.4, it is evident that both the positive signal (Lanes 2–4, 6–8 and 10–12) and background leakage (Lanes 1, 5 and 9) had similar profiles. There was no noticeable circuit leakage even at the high NaCl concentration of 750 mM NaCl.

5.2.3.2 Magnesium Chloride ($MgCl_2$) Concentration

The basic base buffer system was fixed as 1× PBS (pH 7.4) while spiking three concentrations of $MgCl_2$, i.e. 5 mM [20], 10 mM [21] and 12.5 mM [22], to evaluate the effect of $MgCl_2$ concentration. Good signal development was again observed with no noticeable circuit leakage even at the high $MgCl_2$ concentration of 12.5 mM (Fig. 5.5).

5.2.3.3 Reaction Temperature

500 nM of hairpins was reacted with the respective target concentrations in 5× SSCT hybridization buffer at three different temperatures of 4, 25 and 37 °C. Higher temperature outside of the physiologically relevant range was not considered. From Fig. 5.6, the generation of positive signal was reduced slightly when the temperature was lowered. This was most obvious for the case of 0.01 × trigger concentration (Lane 2 vs. 6). Slight circuit leakage was observed at 37 °C (Lane 9), though more positive signal developed as well (Lane 10 vs. 6). The net effect was an improved S/B ratio (Lane 10 over 9), which was still a favourable outcome for detection works.

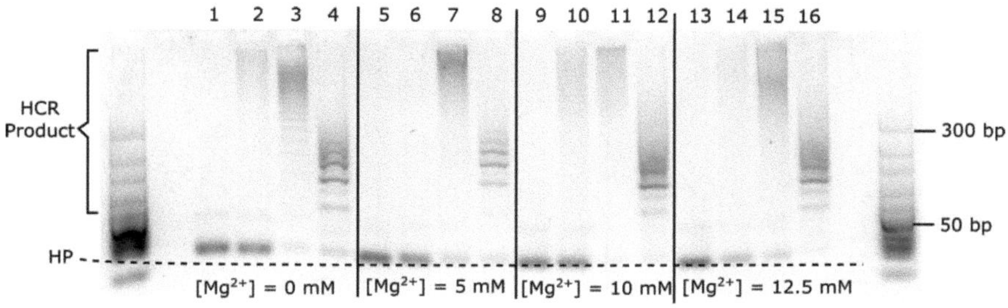

Fig. 5.5 Gel image demonstrating the effect of $MgCl_2$ concentration on the HCR hairpin metastability. 1X PBS (pH 7.4) spiked with varying $MgCl_2$ concentration commonly encountered in HCR studies were used: 0 mM $MgCl_2$ (Lanes 1–4), 5.0 mM $MgCl_2$ (Lanes 5–8), 10.0 mM $MgCl_2$ (Lanes 9–12) and 12.5 mM $MgCl_2$ (Lanes 13–16). For each buffer system, trigger concentration of (from left to right) 0×, 0.01×, 0.1× and 1.0× that of 500 nM hairpins was used. The reaction mixture was incubated at 25 °C for 1 h. A 10–300 bp DNA ladder is shown on both sides of the gel

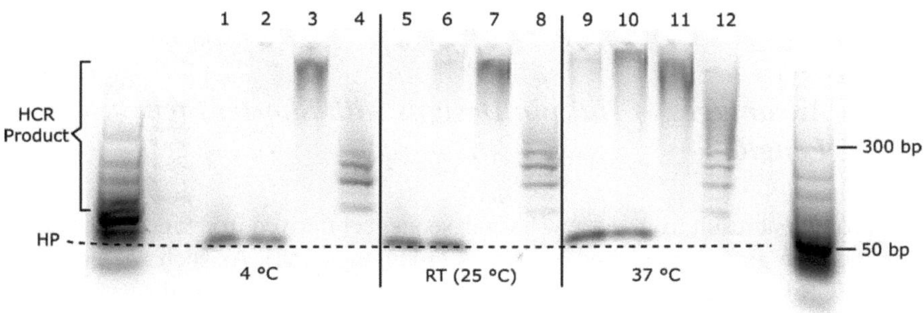

Fig. 5.6 Gel image demonstrating the effect of temperature on the HCR hairpin metastability. We tested 4 °C (Lanes 1–4), room temperature of 25 °C (Lanes 5–8) and 37 °C (Lanes 9–12). For each temperature, trigger concentration of (from left to right) 0×, 0.01×, 0.1× and 1.0× that of 500 nM hairpins was used. The hybridization buffer used was 5× SSCT. The reaction mixture was incubated for 1 h. A 10–300 bp DNA ladder is shown on both sides of the gel

5.2.3.4 Hairpin Concentration

Higher hairpin concentration promotes circuit leakage due to increased occurrence of spurious hybridization. Nonetheless, our hairpins remained metastable in absence of trigger strand (Lane 9) at the highest concentration of 1.0 μM (Fig. 5.7). By comparing the three representative hairpin concentrations of 100 nM (Lanes 1–4), 500 nM (Lanes 5–8) and 1000 nM (Lanes 9–12), we observed that the profile of the HCR product formed was largely dependent on the trigger-to-hairpin ratio rather than the absolute trigger concentration used.

From the above experiments, it was evident that the 6 nt toehold and 12 nt stem design remained robust against leakage over a wide range of experimental conditions, i.e. up to 750 mM NaCl, 12.5 mM $MgCl_2$ and 37 °C for hairpin concentration up to 1.0 μM.

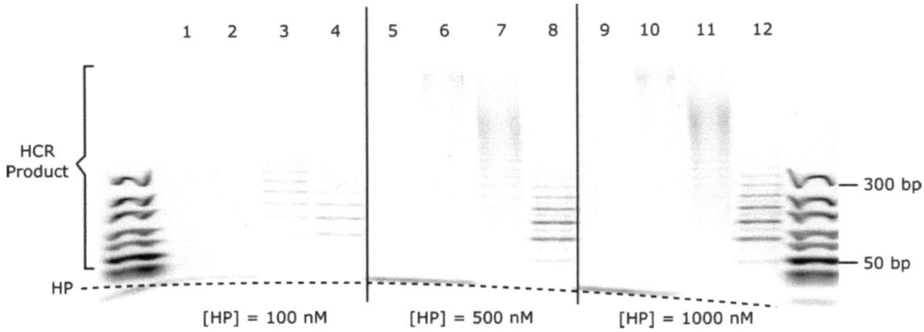

Fig. 5.7 Gel image demonstrating the effect of HCR hairpin (HP) concentration on its metastability. We tested [HP] = 100 nM (Lanes 1–4), [HP] = 500 nM (Lanes 5–8) and [HP] = 1000 nM (Lanes 9–12). For each HP concentration, trigger concentrations of (from left to right) 0×, 0.01×, 0.1× and 1.0× that of the respective HP concentration were used. The hybridization buffer used was 5× SSCT. The reaction mixture was incubated at 25 °C for 1 h. A 10–300 bp DNA ladder is shown on both sides of the gel

5.2.4 *Advantages of Hairpin Designs with Shorter Stem Length*

One prominent result from the new hairpin sequences derived from our design guidelines was the greatly shortened hairpin stem length (12 nt). As such, we anticipate faster HCR kinetics due to the opening of a shorter hairpin stem and shorter branch migration length as compared to the 18 nt stem length in Pierce's design.

It was not possible to evaluate the HCR kinetics using gel electrophoresis directly since it proceeds at a shorter time scale (few mins) than that needed for sample loading and gel running (an hour). A simple method to quench the reaction at different time points was developed. The idea was to terminate the elongation of the HCR product by blocking the single-stranded active domains using complementary protector strands (**P1** and **P2**) (Scheme 5.2). For example, **P1** hybridized to unreacted trigger strand (**T**) to stop further HCR triggering. Also, the growing ends of the HCR product was blocked by **P2** and **P1** at the single-stranded ends of the growing HCR product.

We compared the kinetics of our new sequence design (toehold = 6 nt, stem = 12 nt) and Pierce's original sequence (toehold = 6 nt, stem = 18 nt) using the method presented in Scheme 5.2 [14]. 100 nM of **HP1** and **HP2** reacted with 10 nM of **T** which were quenched at respective time points using two times molar excess (200 nM) of **P1** and **P2**. Two key observations were made, i.e. the intensity of the hairpin bands diminished while the average length of the HCR product increased as the reaction progressed (Fig. 5.8).

The intensities of the hairpin bands were quantified using ImageJ and used to calculate the extent of reaction (defined in Sect. 3.8.1) [23]. The kinetics data were fitted using a pseudo first-order reaction kinetics since the hairpin concentration (100 nM) was greater than **T** (10 nM) (Fig. 5.9). This simple data fitting did not

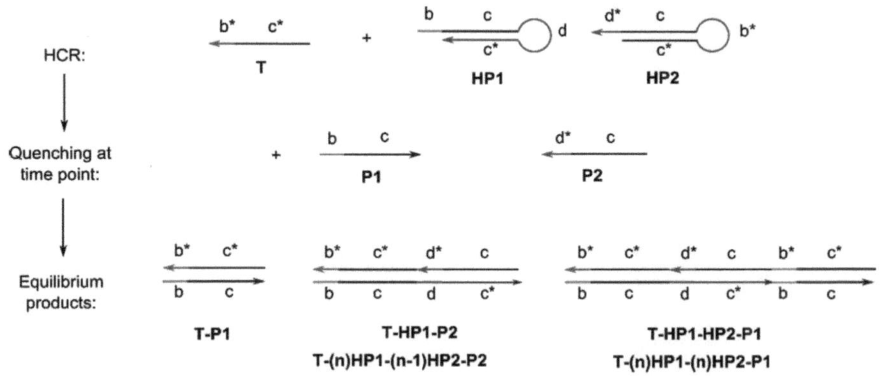

Scheme 5.2 HCR was quenched at different time points using protector strands (**P1** and **P2**) to evaluate its kinetics. These strands were designed to be complementary to the active domains of the trigger strand and growing HCR product. N denotes higher order chain lengths

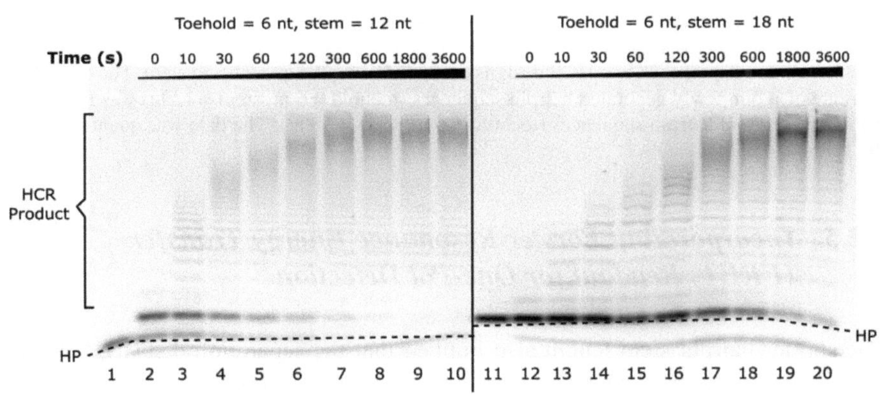

Fig. 5.8 Evolution of the HCR product over time for our hairpin design (Lanes 1–10) and Pierce's design (Lanes 11–20). Lanes 1 and 11 represent 100 nM **HP1** and **HP2**. Lanes 2 and 12 represent 100 nM **HP1** + **HP2** + 200 nM **P1** + **P2**. The reaction mixture (100 nM **HP1** + **HP2** + 10 nM **T**), represented by Lanes 3–10 and Lanes 13–20, was quenched at the time point indicated on top of the respective lanes

fully capture all the complex kinetics involved in HCR and the competing **P1** and **P2** strands, but was adequate for making a semi-quantitative comparison between the two designs. The apparent hybridization rate constant (k) for our sequence was $1.35 \times 10^5\ M^{-1}\ s^{-1}$ which was approximately 10 times faster than that of Pierce's sequence of $1.20 \times 10^4\ M^{-1}\ s^{-1}$. Due to the shorter stem length, less time was required to open the less metastable hairpins and for strand displacement to proceed through the shorter branch migration length.

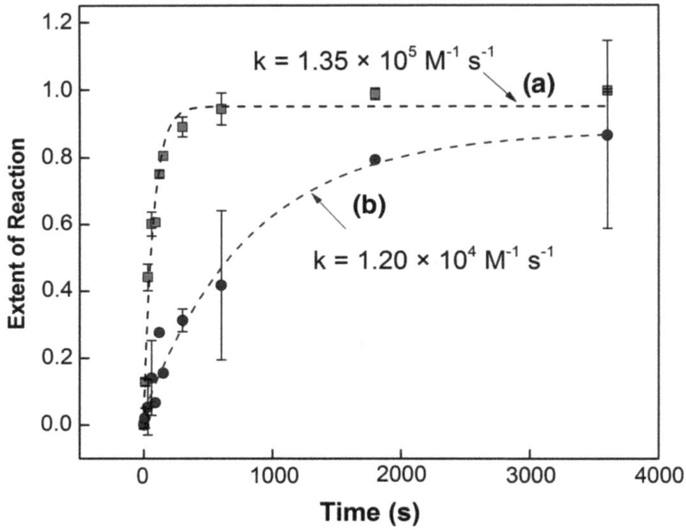

Fig. 5.9 Kinetics profile of two HCR designs evaluated using the quencher system. The hybridization rate of **a** our newly generated hairpin sequences (toehold = 6 nt, stem = 12 nt) was faster than **b** Pierce's original hairpin sequences (toehold = 6 nt, stem = 18 nt). The data was quantified from the gel image in Fig. 5.8

5.2.5 Incorporating Förster Resonance Energy Transfer (FRET) Readout for One-Pot Detection

The shorter hairpin stem length also implied that the separation distance between the 5′ end of hairpin 1 (**HP1**) and 3′ end of hairpin 2 (**HP2**) was reduced from approximately 8.16 to 6.12 nm in the assembled state (Scheme 5.3). As such, we could incorporate a distance-sensitive FRET readout signal by tagging HP1 and HP2 with Cy3 (**HP1-Cy3**) and Cy5 (**HP2-Cy5**) fluorophore at the 5′ and 3′ end respectively. This homogeneous, one-pot turn-on signal based purely on photophysical property was ideal for evaluating the intrinsic performance of our design guidelines due to the lack of complications from mass transfer limitation or the involvement of other steps, e.g. washing or molecule reconfiguration, that were unrelated to HCR itself.

In the original design with a longer 18 nt stem, the theoretical FRET efficiency achievable was only approximately 10.3% which was approximately three times less efficient than the new evolved design (Appendix E). One key advantage of having FRET readout is its ratiometric nature which minimizes interferences from background noise, e.g. instrument variations and environmental fluctuations, using a calculated FRET ratio. We defined this FRET ratio to be the emission intensity of **Cy5-HP2** to **Cy3-HP1**, after correcting for the background emission of a blank set-up.

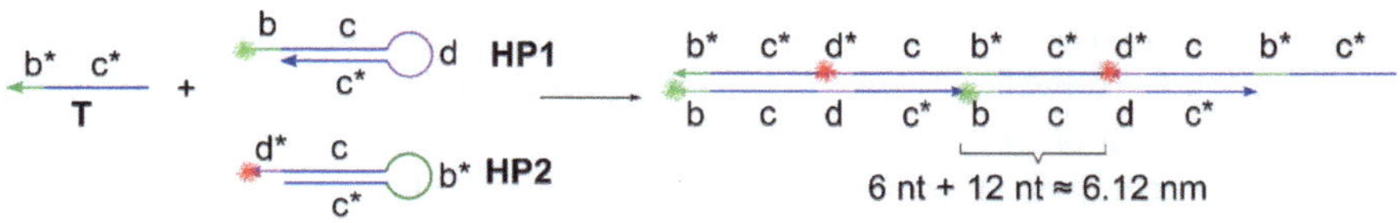

Scheme 5.3 Concept of an HCR Förster resonance energy transfer (FRET) readout system. Hairpin 1 (**HP1**) and hairpin 2 (**HP2**) were labelled with Cy3 and Cy5 fluorophore at their 5′- and 3′-end respectively. Upon binding to the trigger (**T**) strand, HCR occurs. In its assembled state, the two fluorophores were spaced at a constant separation distance of approximately 6.12 nm (combined length of the toehold and stem domain) where efficient energy transfer could still take place

The successfully incorporated FRET readout was used to evaluate the intrinsic performance of HCR as a signal amplification tool. Besides fulfilling its role for upstream characterization, the combined HCR FRET system was further considered as a simple, one-pot detection platform. The use of FRET for localized bio-imaging has been reported in other HCR format although the alternative design is kinetically less favourable than the conventional HCR design [24, 25]. This could be due to the kinetically unfavourable hairpin opening at the loop region instead of the typical linear-linear hybridization in HCR [26].

Our HCR FRET system exhibited reasonably fast kinetics and equilibrium signal was achieved within the timescale of minutes (Fig. 5.10a). The reaction was completed within 60 min even for the slowest kinetics involving the lowest HCR hairpin concentration, i.e. 5.0 nM, used in this study (Fig. 5.10b). Though the initial reaction rate was fastest when equimolar amount of trigger strand was added, the final equilibrium FRET ratio was lower than the next highest trigger concentration. This was due to the formation of incomplete Cy3–Cy5 pairs in shorter HCR assemblies at high trigger concentration [27].

The fluorescence spectrum was analyzed at the 1 h equilibrium mark. As the trigger concentration increased, the Cy3 emission intensity dropped while the Cy5 emission

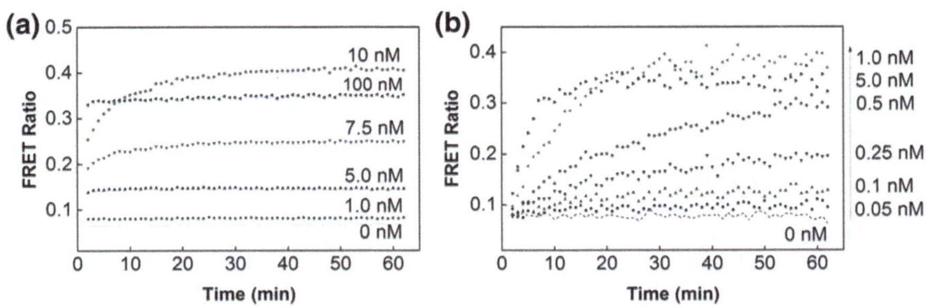

Fig. 5.10 The HCR FRET kinetics was characterized for hairpin concentration of **a** 100 nM and **b** 5.0 nM. Equilibrium was achieved within 60 min even when low hairpin concentration was used. Note that the results were plotted from 2 min onwards to account for the lag time due to mixing and measurement initialization

intensity increased (Fig. 5.11a). The opposite direction of the change in Cy3 and Cy5 emission intensity resulted in distinct increase of the FRET ratio with increasing trigger concentration (Fig. 5.11b). We found interesting trends when varying the hairpin-to-trigger ratio. First, the linear dosage relationship was valid for trigger concentrations below one-tenth that of hairpins. The consistent trigger-to-hairpin relationship suggests that an average of at least 10 pairs of **HP1** and **HP2** should bind to each trigger strand before the saturation effect arising from incomplete Cy3–Cy5 pairs dominates [28]. Compared with high-fold amplification in strategies such as proximity ligation assay [29], the controlled amplification of HCR can be engineered more easily to achieve good sensitivity without signal oversaturation [30].

Next, the limit of detection (LOD), defined to be 3 standard deviations above the background noise, was approximately one-hundredth of the hairpin concentration. This apparent hairpin concentration-dependent LOD threshold was similarly observed by other groups [13, 27], for which the current understanding is that HCR may be terminated by a small fraction of imperfectly-synthesized hairpin. The lowest LOD achieved in this work was 50 pM when 5.0 nM of hairpin was used (Fig. 5.11b).

This performance was comparable to several reported homogeneous HCR assay works in which the LODs are within the picomolar to nanomolar range [12, 13, 31]. We acknowledge that some reported systems have better LOD performance, i.e. femtomolar to low-picomolar range, though it should be noted that they require additional separation or enzymatic amplification steps [1, 7]. Furthermore, the conventional strategy of using fluorescent-labelled HCR hairpins is gaining momentum for biological applications, e.g. fluorescence in situ hybridization study [32], live cell microRNA imaging [25] and flow cytometry analysis [33, 34]. As such, this simple labelling strategy could be of more translational value.

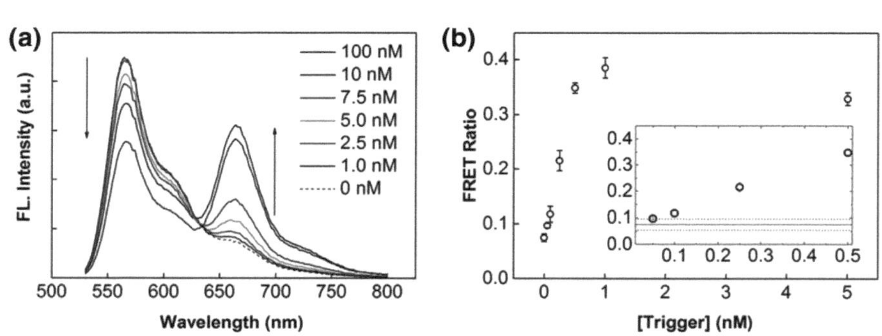

Fig. 5.11 Dosage dependence of the HCR FRET on the trigger concentration. **a** As the trigger concentration increased from 0 to 100 nM, the fluorescence intensity of **Cy3-HP1** decreased while that of **Cy5-HP2** increased. [HP] = 100 nM. **b** The FRET ratio was used to quantify the dosage–signal relationship. The lowest LOD of 50 pM was obtained when [HP] = 5.0 nM. A linear relationship was obtained for the trigger concentration range of 0.05–0.5 nM (R^2 = 0.998). The solid red line corresponds to the mean background noise while the dotted red line corresponds to 3 S.D. from the mean noise level. All data are shown as mean ± S.D. (n = 3). All reactions and characterization were performed at 25 °C

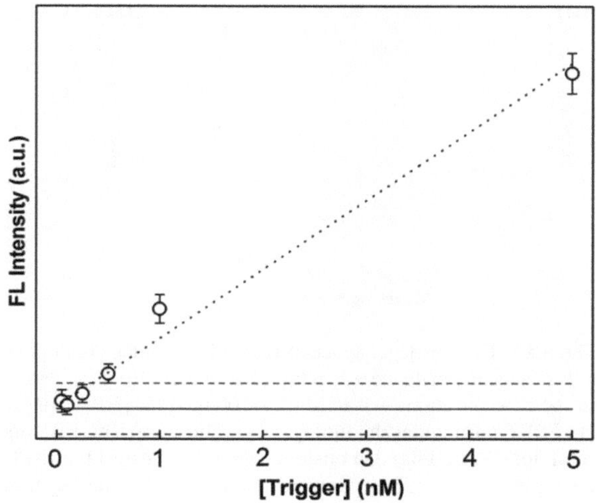

Fig. 5.12 Dosage dependence of the fluorophore-quencher readout on the trigger concentration (0.05–5.0 nM). The solid red line corresponds to the mean background noise while the dotted red line corresponds to 3 S.D. from the mean noise level. All data are shown as mean ± S.D. (n = 3). All reactions and characterization were performed at 25 °C

Compared to a simple fluorophore-quencher (**F-Q**) system, the HCR FRET signal was more robust and had less sample-to-sample variations due to its ratiometric nature. The LOD obtained for a 5.0 nM **F-Q** system was 400 pM (Fig. 5.12), which was about an order of magnitude worse than the HCR FRET system (LOD of 50 pM). This was due to the much higher signal fluctuations and hence larger uncertainties in the background noise.

One notable work from Tan's group has demonstrated similar proximity-based fluorescent readout signal via the formation of pyrene excimer [35]. The pyrene HCR readout has high sensitivity in the femtomolar range without the need for separation steps or secondary amplification. However, the impressive performance under buffered condition could not be replicated in more complex biological environment due to interferences from the high background fluorescence. To ensure that the HCR FRET readout remained a biologically-relevant detection tool, we challenged our system to a biological environment consisting of 10% fetal bovine serum (FBS) at a reaction temperature of 37 °C (Fig. 5.13a). The evolution of fluorescent signals in both buffered and biological environments were similar, thereby demonstrating the robustness of our HCR FRET system to perturbations in the environment (Fig. 5.13b).

5.2.6 Improving the Labelling Schemes of HCR FRET

In our first iteration of the HCR FRET design, the hairpins were labelled at the respective ends to demonstrate the implication of a shorter hairpin stem. It was also due to a technical consideration that the internal fluorophore modification available commercially involves the insertion of fluorophore into the sugar-phosphate DNA background. The effect is akin to a frameshift mutation within the oligonucleotide

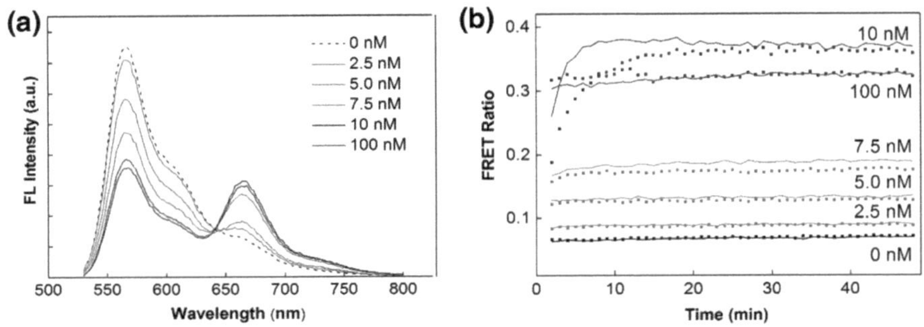

Fig. 5.13 Dosage dependence of HCR FRET under a biological environment of 10% FBS at 37 °C. **a** Its dosage dependence was retained as evident from the fluorescence spectrum. A range of trigger strand concentrations was used (2.5–100 nM) for [HP] = 100 nM. **b** Similar temporal evolution of the FRET ratio under the biological condition of 10% FBS and 37 °C (square dots) and buffered conditions (solid line) demonstrated the robustness of the HCR FRET system. Note that the results were plotted from 2 min onwards to account for the lag time due to mixing and measurement initialization

sequence which will greatly impact the hybridization stability. Considering that the final HCR product is a discontinuous DNA chain, it is possible to label the hairpins strategically without disrupting contiguous stretch of hybridization while further reducing the separation distance between the fluorophores to improve the FRET efficiency.

In the original end-labelled strategy, the fluorophores on the incoming hairpin monomers could pose steric hindrance to the binding of its toehold to the HCR assembly (Scheme 5.4). The effect was non-negligible since the molecular weight of the fluorophores (MW_{Cy3} = 615 Da, MW_{Cy5} = 762 Da) was approximately twice of that of the nucleotides (MW_{nt} = 327). The counter-strategy was to internally conjugate one fluorophore just before the first nucleotide of **HP1** loop, i.e. between domains *c* and *d*, which coincided with the break in hybridization along the HCR assembly. This effectively spaced domains *c* and *d* apart, and relaxed the steric hindrance of the fluorophore on the incoming **HP2**.

Higher FRET signal was obtained for the internally labelled design (IntCy5) compared to the end labelled one when 10 nM of **DT** was added to 100 nM of **HP1** and **HP2** (Fig. 5.14a). Also, the dynamic range of IntCy5 was wider than the end labelled design, where the highest FRET ratio of 5.82 was obtained for 100 nM of **DT** (Fig. 5.14b). Due to near-zero separation distance between Cy3 and Cy5, FRET occurred with maximum efficiency and no longer set an upper bound to the FRET signal attainable.

We next investigated if steric hindrance induced by the physical presence of the larger fluorophores was a significant issue. The larger Cy5 was selected as the internal label since it would space the next hybridization site further away from the as-assembled chain to facilitate the approach of the incoming **HP2** labelled with a smaller Cy3 fluorophore (MW = 615). An additional thymine was inserted before

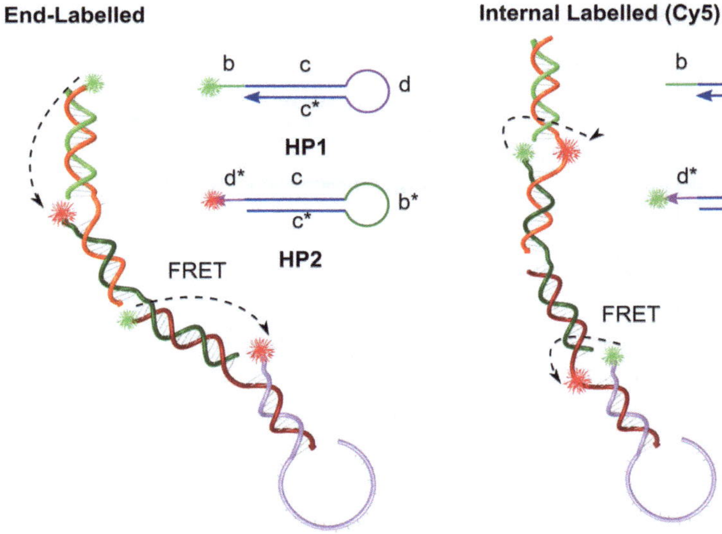

Species	Symbol	Molecular Weight
Nucleotide	-	327
Cy3		615
Cy5		762

Scheme 5.4 Two labelling schemes for the HCR hairpins—both fluorophores labelled at the ends or with Cy5 labelled internally in **HP1**. The helical models are drawn to scale while the "dots-and-lines" monomers are just for illustration

the Cy5 fluorophore in the intCy5 (+1T) design to relax the structure further. The kinetics of signal evolution and the equilibrium signal attained were similar with or without the additional thymine spacer (Fig. 5.14a), suggesting that steric hindrance did not affect the HCR assembly process significantly.

We did notice that the internally modified fluorophore was less photostable and quenched appreciably over time. There were also other complications from the nucleotide identity and the secondary structure of the oligonucleotide on the photostability of the fluorophore [28]. Further optimization works need to be done along this direction. One possible improvement is to use more photostable fluorophores such as the Alexa Fluor series in future iterations of the labelling design.

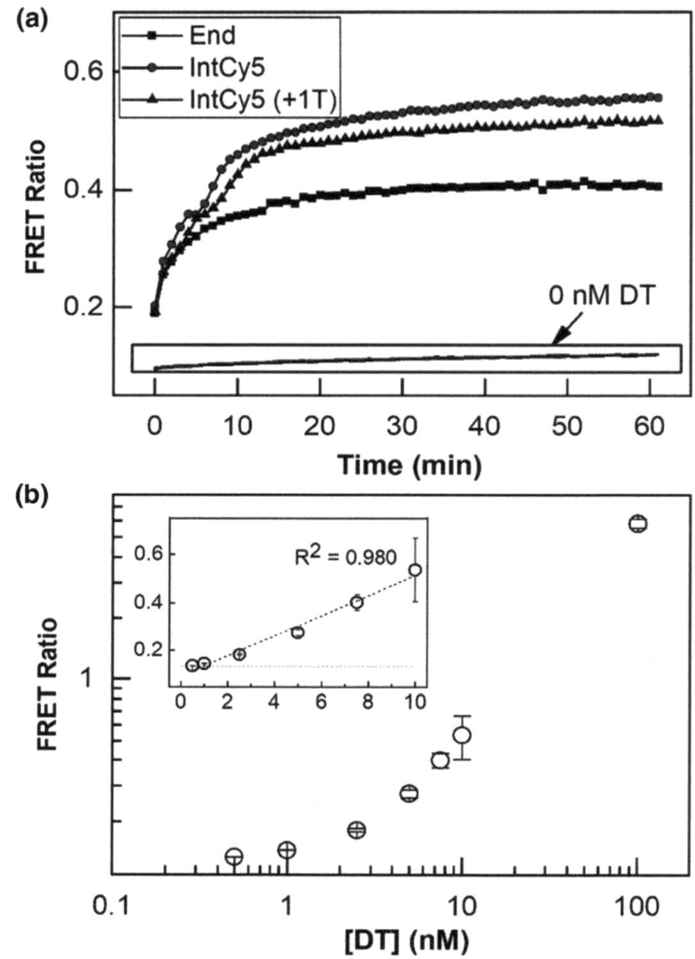

Fig. 5.14 Evaluation of three HCR FRET labelling schemes: end-labelled, internally labelled (IntCy5) and addition of one thymine spacer (IntCy5 (+1T)). **a** Evolution of FRET signal when 10 nM of direct trigger (**DT**) was added to a reaction mixture of 100 nM **HP1** and **HP2**. **b** FRET ratio obtained when different concentration of **DT** (0–100 nM) was added to 100 nM **HP1** and **HP2** for the IntCy5 system. The red solid line represents the mean background noise in absence of the trigger strand while the dashed lines represent ±3 s of this mean. All data are shown as mean ± standard deviation (n = 3). All reactions and characterization were performed at 25 °C

5.3 Conclusion

In conclusion, we have established a set of four-point guidelines to design robust HCR hairpins in terms of the length and C·G content of the toehold and stem domains. The new pair of hairpin sequences generated using this framework were of shorter stem length which improved the HCR kinetics significantly. Next, we successfully incorporated FRET as the readout signal for real-time, one-pot detection. The HCR FRET system exhibited rapid kinetics, tuneable detection range and performed robustly in a biological environment. The signal amplification module was now ready for integration in our overall circuit.

More generally, we hope that the hairpin design guidelines introduced in this chapter can reduce the barrier of using HCR as a viable signal amplification tool, especially for those outside of the DNA circuit field. While more creative and elaborate readout signals have been reported for HCR, we are of opinion that some of these strategies are neither transferrable across reported works nor translatable to actual biologically-relevant applications. FRET, being a commonly-used and well-characterised biological toolbox, should be familiar to a wider audience and encourage greater utilization of our platform.

References

1. Niu, S., Jiang, Y., Zhang, S.: Fluorescence detection for DNA using hybridization chain reaction with enzyme-amplification. Chem. Commun. **46**, 3089 (2010)
2. Zhang, B., Liu, B., Tang, D., Niessner, R., Chen, G., Knopp, D.: DNA-based hybridization chain reaction for amplified bioelectronic signal and ultrasensitive detection of proteins. Anal. Chem. **84**, 5392 (2012)
3. Choi, J., Routenberg Love, K., Gong, Y., Gierahn, T.M., Love, J.C.: Immuno-hybridization chain reaction for enhancing detection of individual cytokine-secreting human peripheral mononuclear cells. Anal. Chem. **83**, 6890 (2011)
4. Chen, X., Briggs, N., McLain, J.R., Ellington, A.D.: Stacking nonenzymatic circuits for high signal gain. Proc. Natl. Acad. Sci. U.S.A. **110**, 5386 (2013)
5. Jiang, Y.S., Bhadra, S., Li, B., Ellington, A.D.: Mismatches improve the performance of strand-displacement nucleic acid circuits. Angew. Chem. Int. Ed. **53**, 1845 (2014)
6. Choi, H.M.T., Beck, V.A., Pierce, N.A.: Next-generation in situ hybridization chain reaction: higher gain, lower cost, greater durability. ACS Nano **8**, 4284 (2014)
7. Wang, X., Lau, C., Kai, M., Lu, J.: Hybridization chain reaction-based instantaneous derivatization technology for chemiluminescence detection of specific DNA sequences. Analyst **138**, 2691 (2013)
8. Yang, X., Yu, Y., Gao, Z.: A highly sensitive plasmonic DNA assay based on triangular silver nanoprism etching. ACS Nano **8**, 4902 (2014)
9. Zheng, J., Hu, Y., Bai, J., Ma, C., Li, J., Li, Y., Shi, M., Tan, W., Yang, R.: Universal surface-enhanced raman scattering amplification detector for ultrasensitive detection of multiple target analytes. Anal. Chem. **86**, 2205 (2014)
10. Dai, S., Xue, Q., Zhu, J., Ding, Y., Jiang, W., Wang, L.: An ultrasensitive fluorescence assay for protein detection by hybridization chain reaction-based DNA nanotags. Biosens. Bioelectron. **51**, 421 (2014)
11. Liu, Y., Luo, M., Yan, J., Xiang, X., Ji, X., Zhou, G., He, Z.: An ultrasensitive biosensor for DNA detection based on hybridization chain reaction coupled with the efficient quenching of a ruthenium complex to CdTe quantum dots. Chem. Commun. **49**, 7424 (2013)
12. Dong, J., Cui, X., Deng, Y., Tang, Z.: Amplified detection of nucleic acid by G-quadruplex based hybridization chain reaction. Biosens. Bioelectron. **38**, 258 (2012)

13. Liu, P., Yang, X., Sun, S., Wang, Q., Wang, K., Huang, J., Liu, J., He, L.: Enzyme-free colorimetric detection of DNA by using gold nanoparticles and hybridization chain reaction amplification. Anal. Chem. **85**, 7689 (2013)
14. Dirks, R.M., Pierce, N.A.: Triggered amplification by hybridization chain reaction. Proc. Natl. Acad. Sci. U.S.A. **101**, 15275 (2004)
15. Tsourkas, A., Behlke, M.A., Bao, G.: Structure-function relationships of shared-stem and conventional molecular beacons. Nucleic Acids Res. **30**, 4208 (2002)
16. Chen, X., Briggs, N., McLain, J.R., Ellington, A.D.: Stacking nonenzymatic circuits for high signal gain. Proc. Natl. Acad. Sci. U. S. A. **110**, 5386 (2013)
17. Zadeh, J.N., Steenberg, C.D., Bois, J.S., Wolfe, B.R., Pierce, M.B., Khan, A.R., Dirks, R.M., Pierce, N.A.: NUPACK: analysis and design of nucleic acid systems. J. Comput. Chem. **32**, 170 (2011)
18. Koos, B., Cane, G., Grannas, K., Lof, L., Arngarden, L., Heldin, J., Clausson, C.-M., Klaesson, A., Hirvonen, M.K., de Oliveira, F.M.S., Talibov, V.O., Pham, N.T., Auer, M., Danielson, U.H., Haybaeck, J., Kamali-Moghaddam, M., Soderberg, O.: Proximity-dependent initiation of hybridization chain reaction. Nat. Commun. **6**, 7294 (2015)
19. Chen, X.: Expanding the rule set of DNA circuitry with associative toehold activation. J. Am. Chem. Soc. **134**, 263 (2012)
20. You, M., Peng, L., Shao, N., Zhang, L., Qiu, L., Cui, C., Tan, W.: DNA “nano-claw”: logic-based autonomous cancer targeting and therapy. J. Am. Chem. Soc. **136**, 1256 (2014)
21. Tang, Y., Wang, Z., Yang, X., Chen, J., Liu, L., Zhao, W., Le, X.C., Li, F.: Constructing real-time, wash-free, and reiterative sensors for cell surface proteins using binding-induced dynamic DNA assembly. Chem. Sci. **6**, 5729 (2015)
22. Wu, C., Cansiz, S., Zhang, L., Teng, I.T., Qiu, L., Li, J., Liu, Y., Zhou, C., Hu, R., Zhang, T., Cui, C., Cui, L., Tan, W.: A nonenzymatic hairpin DNA cascade reaction provides high signal gain of mRNA imaging inside live cells. J. Am. Chem. Soc. **137**, 4900 (2015)
23. Schneider, C.A., Rasband, W.S., Eliceiri, K.W.: NIH Image to ImageJ: 25 years of image analysis. Nat. Methods **9**, 671 (2012)
24. Wu, Z., Liu, G.-Q., Yang, X.-L., Jiang, J.-H.: Electrostatic nucleic acid nanoassembly enables hybridization chain reaction in living cells for ultrasensitive mRNA imaging. J. Am. Chem. Soc. **137**, 6829 (2015)
25. Cheglakov, Z., Cronin, T.M., He, C., Weizmann, Y.: Live cell microRNA imaging using cascade hybridization reaction. J. Am. Chem. Soc. **137**, 6116 (2015)
26. Tsourkas, A., Behlke, M.A., Rose, S.D., Bao, G.: Hybridization kinetics and thermodynamics of molecular beacons. Nucleic Acids Res. **31**, 1319 (2003)
27. Li, B., Jiang, Y., Chen, X., Ellington, A.D.: Probing spatial organization of DNA strands using enzyme-free hairpin assembly circuits. J. Am. Chem. Soc. **134**, 13918 (2012)
28. Nazarenko, I., Pires, R., Lowe, B., Obaidy, M., Rashtchian, A.: Effect of primary and secondary structure of oligodeoxyribonucleotides on the fluorescent properties of conjugated dyes. Nucleic Acids Res. **30**, 2089 (2002)
29. Fredriksson, S., Gullberg, M., Jarvius, J., Olsson, C., Pietras, K., Gustafsdottir, S.M., Ostman, A., Landegren, U.: Protein detection using proximity-dependent DNA ligation assays. Nat. Biotechnol. **20**, 473 (2002)
30. Mocanu, M.-M., Váradi, T., Szöllősi, J., Nagy, P.: Comparative analysis of Fluorescence Resonance Energy Transfer (FRET) and Proximity Ligation Assay (PLA). Proteomics **11**, 2063 (2011)
31. Ma, C., Wang, W., Mulchandani, A., Shi, C.: A simple colorimetric DNA detection by target-induced hybridization chain reaction for isothermal signal amplification. Anal. Biochem. **457**, 19 (2014)
32. Choi, H.M.T., Chang, J.Y., Trinh, L.A., Padilla, J.E., Fraser, S.E., Pierce, N.A.: Programmable in situ amplification for multiplexed imaging of mRNA expression. Nat. Biotechnol. **28**, 1208 (2010)

33. Ren, W., Liu, H., Yang, W., Fan, Y., Yang, L., Wang, Y., Liu, C., Li, Z.: A cytometric bead assay for sensitive DNA detection based on enzyme-free signal amplification of hybridization chain reaction. Biosens. Bioelectron. **49**, 380 (2013)
34. Zhang, Y., Liu, C., Sun, S., Tang, Y., Li, Z.: Phosphorylation-induced hybridization chain reaction on beads: an ultrasensitive flow cytometric assay for the detection of T4 polynucleotide kinase activity. Chem. Commun. **51**, 5832 (2015)
35. Huang, J., Wu, Y., Chen, Y., Zhu, Z., Yang, X., Yang, C.J., Wang, K., Tan, W.: Pyrene-excimer probes based on the hybridization chain reaction for the detection of nucleic acids in complex biological fluids. Angew. Chem. Int. Ed. **50**, 401 (2011)

Chapter 6
Design Concepts in Association Toehold for Robust Signal Transduction

6.1 Introduction

We have shown that aptamer-based molecular switch, which re-configures upon target binding, can be used to trigger signal transduction in our self-contained tool in Chap. 4. While this is an elegant strategy, limited number of well-validated aptamer sequences are available as compared to more conventional recognition moieties such as antibody [1]. From this limited aptamer pool, not all of them can be easily designed as a target-responsive molecule switch which is dependent on the type of molecular reconfiguration induced from target binding.

This limitation motivates the design of a more general signal transduction strategy to detect a wider pool of biomolecule targets. We aim to evolve the self-contained circuit into one that is triggered by the reassembly of split circuit components when their respective targets interact. An analogy is the split-protein reporter system based on the reassembly of the readout signals, typically fluorescence protein, from the binding event between the tagged biomolecules [2, 3]. This simple principle is general and can readily be extended to synthetic, cell-free systems as demonstrated by Ghosh and co-workers using protein reporters synthesized from cell-free translation [4].

The equivalent idea in DNA circuit design is to split the trigger strand. Previous study on remote toehold suggests that toehold mediated strand displacement can still take place across non-adjacent domains, though incurring penalty on the kinetics [5]. In a separate work, Chen demonstrated the concept of "associative toehold activation" where the toehold and branch migration domains are decoupled on separate DNA strands and later activated to reassemble in response to specific hybridization events [6]. In this work, we will demonstrate that the reassembly process can be triggered by physical events which brings the split domains into close proximity besides the usual DNA hybridization.

Our strategy is to split a single reporter trigger strand (c^*b^*) and attach the individual domain on separate initiator strands (**I1** and **I2**) (Scheme 6.1). The initiator strands bind to the respective targets and if their targets are interacting in close proximity,

A. Yan Shan, *Engineering a Robust DNA Circuit for the Direct Detection of Biomolecular Interactions*, Springer Theses,
https://doi.org/10.1007/978-981-13-2188-7_6

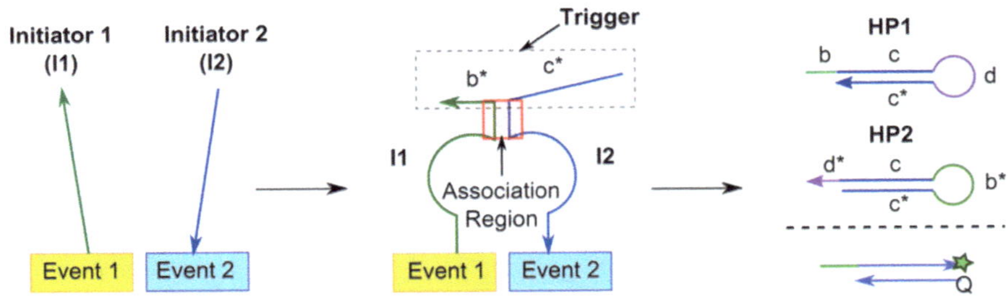

Scheme 6.1 Association toehold as the signal transduction module. The circuit design involved three steps: (1) binding of initiator 1 (**I1**) and initiator 2 (**I2**) strands to the respective recognition sites. (2) The close proximity of **I1** and **I2** kinetically favoured the formation of a complete trigger strand ($c^*\ b^*$) which was stabilized by a short DNA association region to (3) initiate the readout signals

the local concentration of the initiators increases and promotes the reassembly of the split domains into a complete trigger sequence. Similar strategy was demonstrated by Le and co-workers as a wash-free method for endpoint protein detection [7–9].

An association region is included to thermodynamically stabilize the intermediate **I1-I2** complex (Scheme 6.2) [6]. The association region should be short (approximately 4–6 nt) so that they do not form stable complexes when diffusing randomly in solution. The more formal criterion is to keep its melting temperature (T_m) way below room temperature; T_M for 4–6 nt (at 50% C·G content) was found to be below 10 °C using mfold web server simulation [10]. Upon recognizing a pair of interacting targets, the initiators are localized within the confines of the targets. Hybridization of the association region is no longer about two DNA strands randomly coming into contact in free solution. Instead it is the probability of the two strands meeting onto another within the bound complex. This construct is similar to the hairpin loop structure where a short stem length suffices in stabilizing the structure drastically, i.e. T_M increases to the range of 40–50 °C.

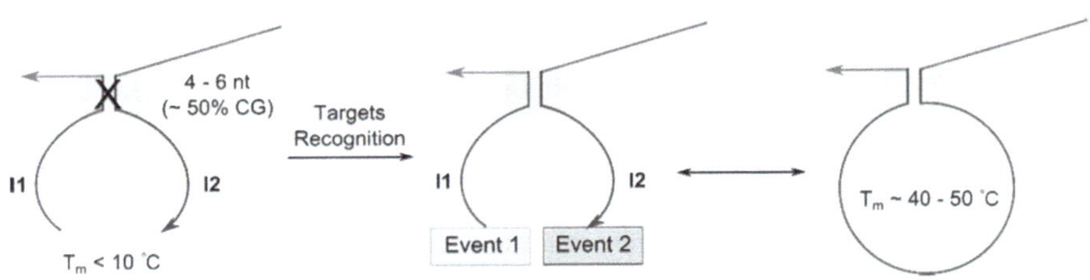

Scheme 6.2 The stability of the association toehold on its interacting target(s) approached that of a molecular hairpin. The association region on the initiators (**I1** and **I2**) is kept short (4–6 nt) so that the melting temperature (T_m) of the duplex is less than 10 °C. Upon target recognition, the initiators are localized in close proximity which resembles that of a hairpin structure where a short stem length drastically stabilizes the secondary structure (T_m increases to approximately 40–50 °C)

The critical role of the association region (domain *a*) was proven experimentally where the lack of domain *a* slowed down the kinetics considerably and the signals obtained after an hour of reaction was only approximately one-fifth of that when a short 6 nt domain *a* was included (Fig. 6.1). We will refer this signal transduction mechanism as "association toehold" from this point onwards.

The performance of any DNA circuit is closely related to its signal-to-background (S/B) ratio. We broadly classify all background noise due to any undesired circuit operation as circuit leakages. In our case, this refers to any signal that is not localized at the target binding site, or did not result from two targets in close proximity. The issue of circuit leakage is of concern in our association toehold system as the critical domains (domains *a, b** and *c**) are all exposed as active, single stranded form, unlike the traditional approach where they are locked up as duplexes. As such, the central aim of this chapter is to identify the sources of leakages involved and systematically formulate design guidelines to minimize the circuit leakages.

We had previously used low probe concentrations to kinetically suppress undesirable leakages in Chap. 4, which is an approach also used by other groups [8, 11]. While it may be effective within the scope of a specific study, this approach lacks generality and may require arbitrary re-optimization for individual reaction conditions. Alternatively, a design-driven approach based on careful thermodynamics considerations can guarantee a more robust outcome and will be the method of choice used in this chapter wherever possible [12, 13].

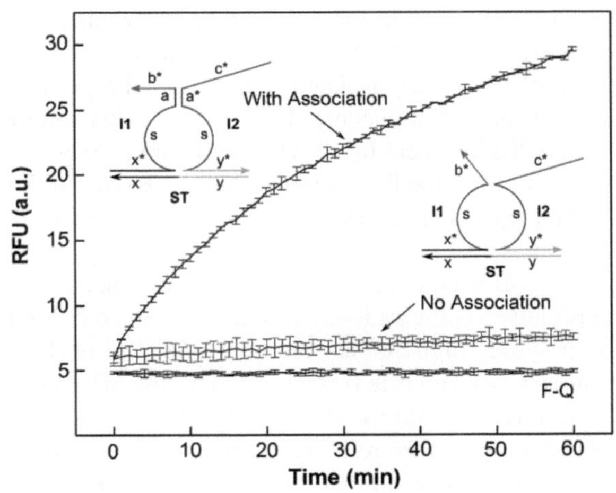

Fig. 6.1 The association region (domain *a*) was necessary to stabilize the split trigger (**ST**) domains (c^* d^*) for signal generation. The length of domain *a* was 6 nt here. All data is presented as mean ± standard deviation (n = 3)

6.2 Results and Discussion

6.2.1 *Sources of Circuit Leakage in Association Toehold System*

We used the general framework formalized by Ellington's group to classify the types of leakages experienced in our association toehold system [14]. The two types of leakages are: (1) initial leakage due to oligonucleotide synthesis defects and misfolded DNA, and (2) asymptotic leakage due to spurious hybridization, conformational fluctuations or cross-talk between domains [14]. The former may be minimized, though not eliminated, by using purified (HPLC- or PAGE- grade) oligonucleotides. The latter is inherent at the point of sequence design and requires deliberate design effort to systematically minimize background noise for any circuit system.

The operation of the association toehold was demonstrated on an ideal system consisting of a single-stranded DNA split target (**ST**) strand with two recognition domains (domains x and y) to mimic bi-molecular binding events (Fig. 6.2a). The circuit was evaluated at various design stages using the hybridization chain reaction (HCR) [15, 16] and/or fluorescence-quencher (**F-Q**) readout. The former amplified the presence of circuit leakages and when used in conjunction with gel electrophoresis, allowed for the unambiguous identification of various DNA complexes. The latter was used to profile the kinetics evolution of various leakage types.

In our first design iteration of the association toehold, we used 6 nt for the association region and 9 nt for domain b^* as previously reported by Le's group [8]. There were two key sources of leakage in the association toehold system (Fig. 6.2a), which were systematically isolated and confirmed experimentally by the incremental addition of DNA components (Fig. 6.2b). Leak I referred to the toehold-independent strand displacement by **I2**, which domain c^* was available to displace the identical **Q** strand from the fluorophore-labelled (**F**) strand. Leak II referred to the formation of **I1-F-Q** and **I1-I2** intermediate complexes when all DNA components, except **ST**, were present.

Leak II accounted for the bulk of the leakage which was not out of expectation especially via the pathway of intermediate complex I. However, its magnitude was surprisingly high and accounted for 56.57% of the positive signal despite using a similar format reported previously. This might be due to the significantly higher probe concentration (100 vs. 10 nM) or different reaction buffer used in our study. This observation further reinforced our argument that kinetically-suppressed leakages lacked robustness. In fact, Leak II is an inherent feature of the association toehold design which will be further discussed in Sect. 6.2.4. On the other hand, Leak I contributed to a smaller but non-negligible amount of background noise. Nonetheless, its effect could be amplified when used in conjunction with our HCR FRET readout.

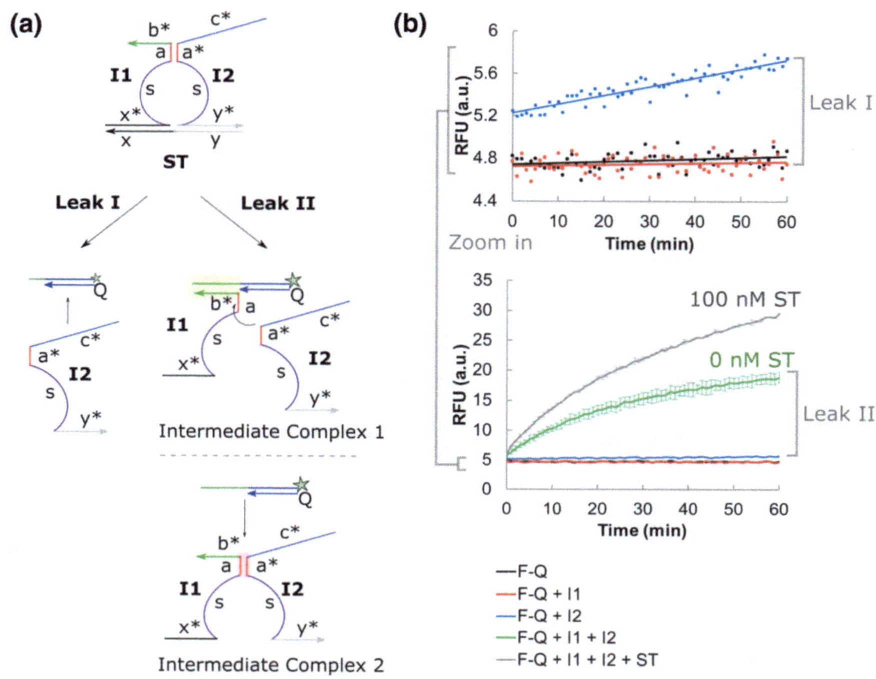

Fig. 6.2 Sources of leakages in the proposed association toehold system. **a** There were two main sources of circuit leakage: Leak I—toehold-independent strand displacement by initiator 2 (**I2**), and Leak II—formation of intermediate complex 1 (**I1-F-Q**) or 2 (**I1-I2**) in presence of all DNA components except **ST**. **b** Different combinations of the DNA components (**F-Q**, **I1**, **I2** and **ST**) revealed the types of leakages involved. Error bars shown indicate sample standard deviation of triplicate experiments. However, error bars were not shown for the narrow RFU range (top graph showing Leak I) to avoid congesting the plot

6.2.2 3′-Truncated Oligonucleotide as the Source of Initial Leakage

Leak I could be further decoupled into two types of leakages, i.e. initial leakage which was characterized by an initial signal burst and asymptotic leakage which steadily grew with time (Fig. 6.3). We first tackled the issue of initial leakage which was caused by synthesis defects in the oligonucleotide. Synthesis defect refers to the failed coupling of nucleotides at various points of the stepwise synthesis process, resulting in a mixture of truncated sequences.

We deduced that such defects on **Q** strand was the main culprit of the initial leakage (refer to Scheme 6.3 for the rationale of this hypothesis) since they left gaps on the **F-Q** complex from which **I2** could initiate hybridization with **F**. A possible counter-strategy was to shorten **I2** to thermodynamically disfavour the undesired

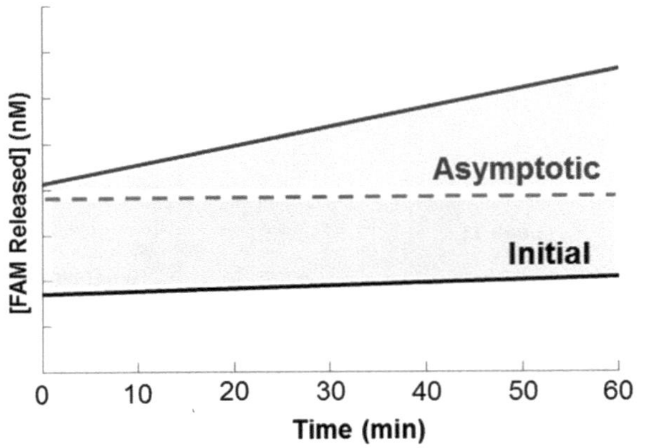

Fig. 6.3 Leak I could be further decoupled into types of leakages, i.e. initial leakage and asymptotic leakage

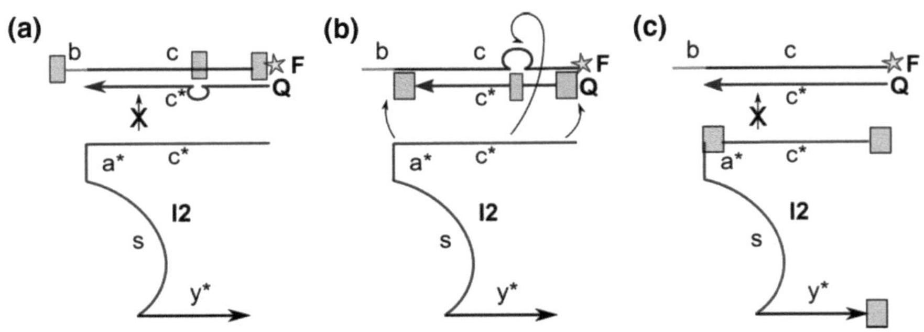

Scheme 6.3 Identification of the oligonucleotide defect site contributing the most to initial leakage. Three sites of synthesis defect (3′, 5′ and internal), as indicated by grey boxes, are possible for the n-1 oligonucleotide product of **a** fluorophore (**F**) strand, **b** quencher (**Q**) strand and **c** initiator 2 (**I2**) strand. By logical deduction, the presence of defects on **F** and **I2** cannot release the reporter signal. Instead, the defects on **Q** can expose sites on the **F-Q** complex for strand displacement by **I2** to form a more thermodynamically-stable **I2-F** product

displacement of **Q**. This was done by eliminating one nucleotide each from the 5′- and 3′-end of domain $c*$ on **I2** (Fig. 6.4a). Defects in the internal sites were not considered due to the inherent randomness in their position.

The truncation of domain $c*$ at the 3′-end of **I2** (**I2 (-3′)**) resulted in the least signal burst and was most effective in reducing the initial leakage (Fig. 6.4b). This suggests that most gaps were found on the corresponding 3′-end of **Q** which appeared to be more vulnerable to synthesis defects. We also noted that the initial burst from the 5′-truncated **I2** strand was higher than the full length **I2**. The reason for this unexpected observation is unknown, however, the decreasing leakage over time can be explained by the subsequent strand displacement of **I2 (-5′)** by the full-length **Q** strand.

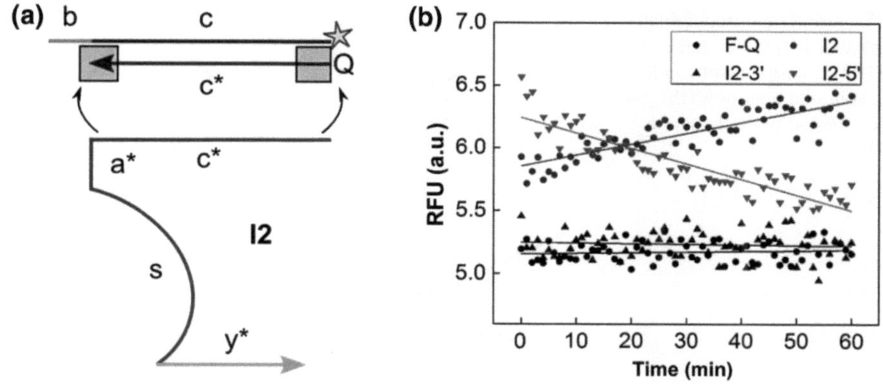

Fig. 6.4 The 3′-end of **Q** was found to be responsible for initial leakage. **a** The corresponding nucleotide on **I2** was removed to isolate which end (3′- or 5′-) of **Q** (grey box) contributed predominantly to the leakage. **b** The omission of a single nucleotide at the 3′-end of **I2** domain *c** (**I2-3′**) was most effective in minimizing the initial leakage. The mean values of triplicate readings were presented as scatter plot, while error bars were not shown for the narrow RFU range to avoid congesting the plot

Scheme 6.4 Solid-phase synthesis of commercially available oligonucleotide proceeds in a 3′- to 5′-direction from a glass bead support

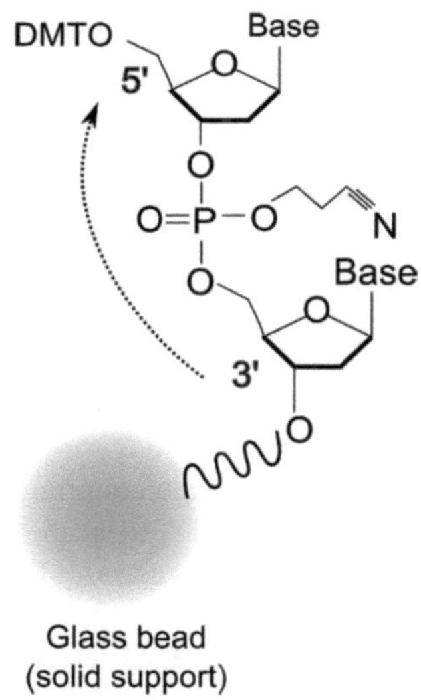

This observation can be explained by considering that the solid-phase oligonucleotide synthesis proceeds in a 3′- to 5′- direction (Scheme 6.4). Hence the first few nucleotides closer to the solid support, i.e. at the 3′ end, experience greater steric hindrance and do not couple efficiently. Our observation was also consistent with an old report by Temsamani et al. where 55% of their n-1 PAGE-purified products were found to be truncated at the first four nucleotides from the 3′-end [17].

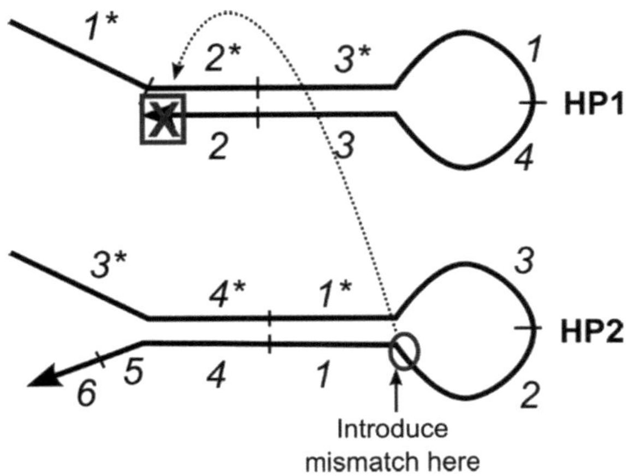

Fig. 6.5 The introduction of mismatch at the 3′-end of **HP2** domain *2* results in less favourable binding to the gap left behind by 3′-defect on **HP1**

However, this contracted the commonly-held belief by several groups working in this area that more defects are located at the 5′-end [13, 18]. We presume that this belief stems from their understanding that more coupling errors accumulate as the oligonucleotide strand grows. While it is true that the synthesis errors are accumulative, i.e. a longer sequence design will result in a greater overall percentage defect, it is inaccurate to think that the bulk of the error occurs only towards the end of the synthesis procedure at the 5′-end of the oligonucleotide.

Previously unexpected experimental results on the effect of DNA impurities and 5′/3′ orientation may be re-interpreted more intuitively in light of our alternative perspective and evidence presented [13]. It can also partially explain why the presence of mismatches in the position corresponding to the 3′-end of the other CHA hairpin pair is most effective in suppressing the CHA leakage (Fig. 6.5) [19].

Oligonucleotides used in DNA nanotechnology works are often obtained via solid-phase synthesis, rendering this a common yet not often discussed issue across the field. The average coupling efficiency of the oligonucleotides used in this work is guaranteed by IDT to be above 99% per reaction for synthesis lengths up to 100 bases. Oligonucleotides used for DNA circuit studies are minimally HPLC purified as in our case; while a higher purity of >90% can be obtained for PAGE purified sequences. As such, the average error rate in these synthetic sequences is about 1–10%. We believe that the added knowledge of this non-trivial fact on the 3′-end being defected can guide future circuit designs or purification methods closer towards achieving an ideal leak-free system.

Alternatively, other synthesis and purification methods may be considered to circumvent this technical issue. These alternative methods often involve biological replications with much lower error rates, typically in the order of 10^{-7}–10^{-8}, [20] such as via enzymatic production of oligonucleotides [14, 21]. Though the end results are remarkable, numerous manipulation steps are involved, rendering these methods more suitable for use in a central facility rather as a routine procedure for any labs. Mass production of high purity oligonucleotides is possible using the technique developed by Högberg and co-workers [22] where defect-free templates from bacteria colony are amplified using rolling circle amplification and the desired sequences are obtained via restriction enzyme digestion. Otherwise, one can try to minimize leakages due to oligonucleotide defects simply by improving the inherent circuit design.

6.2.3 "Inter-domain Bridging" for Improving Strand Displacement Across Split Domains

Since the strand displacement reaction is in dynamic equilibrium, the exposed domain c* of **I2** can displace its identical counterpart, **Q**, over time and contribute to asymptotic leakage in the original design (Fig. 6.6a). We conceptualized three designs in Fig. 6.6a to thermodynamically disfavour the toehold-independent strand displacement experienced in the original design (Design 1). The common feature of the designs is to render the nucleotide at the 3′-end of domain $c*$ inaccessible for strand displacement. In Design 2, the entire domain $c*$ of **I2** was protected with a lock strand (**I2-D2L**). The next two designs focused on changing a single nucleotide at the 3′-end of domain $c*$, i.e. introducing a single mismatch (**I2-D3** in Design 3) or translocating the nucleotide to the 5′-end of domain $b*$ in **I1** (**I2-D4** in Design 4).

All three designs (Designs 2–4) successfully suppressed the asymptotic leakage as seen from the relatively stable background noise over time (Fig. 6.6b). Design 2 was most effective in reducing the initial leakage (87.1% reduction in initial fluorescent burst) compared to the 13.5% and 16.6% reduction by Design 3 and 4 respectively. This was because the nucleotide at the 3′-end of domain $c*$ in Design 2 was completely locked up by the complementary **I2-D2L** strand. However, inactivating the 3′ nucleotide on domain $c*$ could penalize the generation of positive signal as observed in Design 2 due to the unfavourable strand displacement kinetics across a four-way junction (Fig. 6.6c) [23].

For Design 3, the single mismatch introduced at the first incumbent nucleotide of domain $c*$ also hindered the effective initiation of branch migration. Since branch migration is a nucleotide-by-nucleotide random walk process and the initiation of branch migration is thermodynamically unfavourable (Sect. 2.2.2), the approach by the first few nucleotides after the toehold binding is critical in determining if the branch migration continues forward or the bound toehold duplex dissociates [24].

Out of these, the modified **I2-D4** exhibited the best performance in terms of both suppressing the asymptotic leakage, as seen from the relatively stable background noise over time (Fig. 6.6b), and in boosting the generation of positive signal by 65.9% as compared to the original **I2** (Fig. 6.6c). These observations were confirmed using gel electrophoresis in junction with HCR readout to amplify the minute amount of circuit leakage (Fig. 6.6d).

The concept of inter-domain bridging arising from the translocation of nucleotide from domain $c*$ to $b*$ can be used to explain this observation (Scheme 6.5). After

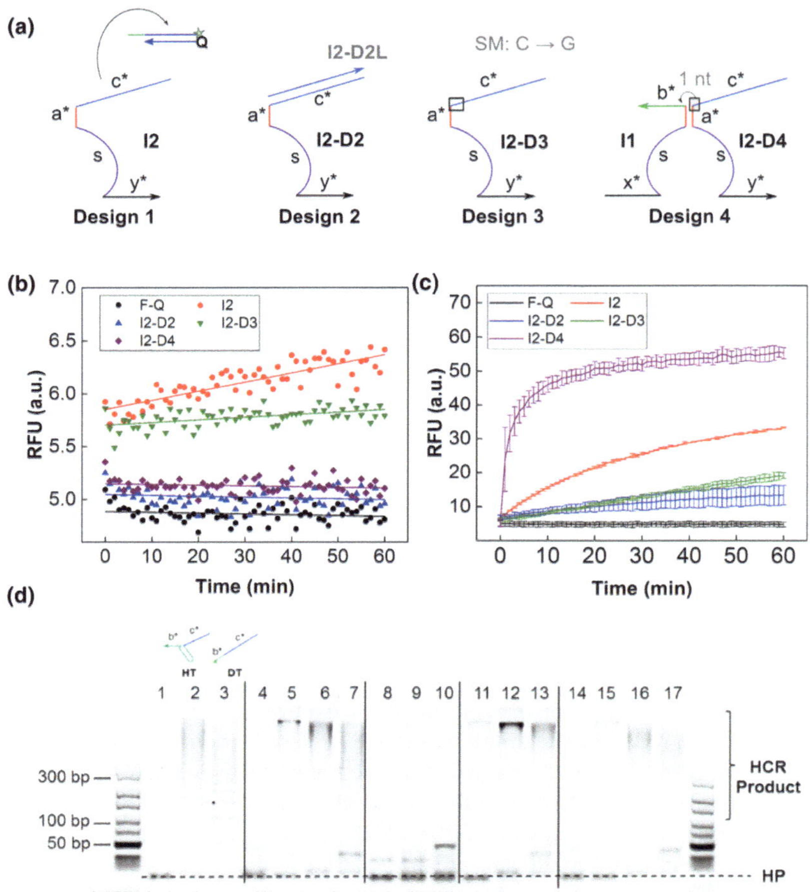

	Controls			Design 1				Design 2			Design 3			Design 4			
HP1+HP2	+	+	+	+	+	+	+	+	+	+	+	+	+	+	+	+	+
I1	-	-	-	+	-	+	+	-	+	+	-	+	+	+	-	+	+
I2	-	-	-	-	+	+	+	+	+	+	+	+	+	-	+	+	+
T	-	HT	DT	-	-	-	ST	-	-	ST	-	-	ST	-	-	-	ST

Fig. 6.6 **a** Three strategies (Designs 2–4) were conceptualized to suppress the toehold-independent strand displacement (Design 1). **b** The generation of Type I leakage over time was evaluated in a **F-Q + I2** reaction mixture. The mean values of triplicate readings were presented as scatter plot, while error bars were not shown for the narrow RFU range to avoid congesting the plot. **c** Time-profile of positive signals generated in the presence of 100 nM **ST** and all circuit components (**F-Q + I1 + I2**) showed that Design 4 generated the highest signal intensity at the fastest rate. Error bars shown indicate sample standard deviation of triplicate experiments. **d** Gel electrophoresis was used to qualitatively confirm the observations. 500 nM of individual DNA components was used. **HT** refers to hairpin trigger, **DT** refers to direct trigger and **ST** refers to split trigger. 10–300 bp DNA ladders are shown at both sides of the gel

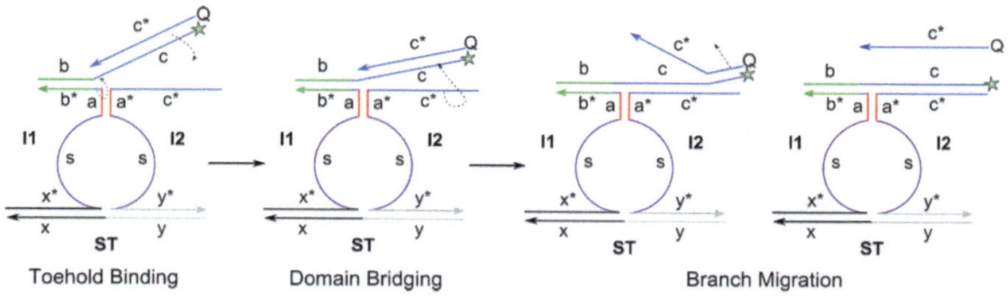

Scheme 6.5 The translocation of one nucleotide from domain $c*$ to domain $b*$ relieved the distinct separation of domains, in a strategy which we termed inter-domain bridging. First, toehold binding took place at domain b. Next, the single nucleotide brought over from domain $c*$ initiated the displacement of the quencher (**Q**) strand and pulled the fluorophore-quencher (**F-Q**) complex towards domain $c*$ of **I2**, effectively bridging the split domains spatially. Finally, strand displacement continued *via* the usual branch migration process

the **F-Q** complex binds to **ST-I1-I2** at the toehold domain, the single translocated nucleotide at the 5′-end of domain $b*$ initiated the branch migration of **Q** (a thermodynamically unfavourable process as discussed in Sect. 2.2.2) and anchored the **F-Q** complex towards domain $c*$ on **I2-D4** strand. The initial reaction rate of the overall strand displacement process increased from $k_{\mathbf{I2\text{-}D1}} = 1.44$ nM min^{-1} to $k_{\mathbf{I2\text{-}D4}} = 18.8$ nM min^{-1}. This suggests that the closer spatial proximity between **F-Q** and **I2** strands promoted the kinetics of the remaining branch migration process. Overall, a two-fold improvement in the signal-to-background (S/B) was achieved simply by shifting a single nucleotide across the split domains (S/B of **I2-D4** = 9.5 vs. S/B of **I2-D1** = 5.2).

6.2.4 Analysis of Intermediate Complexes Contributing to Leak II

Two types of intermediate complexes involved in the formation of the three-way junction were responsible for Leak II (Fig. 6.2a). We assumed that two out of the three strands involved (**I1, I2** and **F-Q**) had to first form a stable intermediate complex for the third incoming strand to bind to and trigger the reporter signal since a simultaneous tri-molecular collision event is rare [25].

I1 and **F** could hybridize at domain $b*$ while the association region (domain $a*$) doubled as the reporter toehold to trigger the formation of the reporter signals (intermediate complex 1). Also, **I1** and **I2** could associate at domain $a*$ in the absence of the target (intermediate complex 2). It was difficult to isolate individual complexes experimentally because both domains $a*$ and $b*$ played dual roles of complex stabilization (for the formation of intermediate complex) and reporter triggering (branch migration to release the reporter signal) (Fig. 6.2a). Regardless, it was possible to deduce the relative contribution by each intermediate complex from their respective thermodynamics stability which was determined by the length of domains a^* and b^* respectively.

In our original configuration ($a^* = 6$ nt and $b^* = 9$ nt), the S/B ratio decreased over time (Fig. 6.7a). This was due to the extensive amount of Leak II whereby the intermediate complexes consumed 39.7 nM of reporter strands. Based on NUPACK simulation, **I1-F-Q** (intermediate complex 1) was found to be the predominant species at equilibrium (45.0 nM) compared to **I1-I2** (intermediate complex 2, 0.61 nM). When the length of domain b^* was reduced to 7 nt, the S/B ratio reversed to an increasing trend (Fig. 6.7a). This was due to the significantly lower equilibrium concentration of **I1-F-Q** complex (2.6 nM) which reduced the Leak II consumption of reporter strands to 3.1 nM. Further reduction of the length of domain b^* to 5 nt greatly penalized the generation of positive signal, leading to a negligibly low S/B ratio (Fig. 6.7a).

From the above result, we propose that domain b^* is the de facto reporter toehold in the proper **ST-I1-I2** assembly and should be kept at a kinetically functional length ($\geq$6 nt) to significantly enhance the triggering of reporter signals in the presence of **ST** [26]. The underlying assumption is that while strand displacement can be triggered by dynamic toehold binding (a non-equilibrium event), the intermediate complexes receiving the incoming strand should be sufficiently stable at equilibrium. This is based on the understanding that a simultaneous three-entity collision leading to meaningful product formation is an extremely rare event.

We consider the case where the length of domain b^* is greater than that of domain a^*, which is now constrained to $\leq$5 nt (Fig. 6.8a). Under this constraint, no **I1-I2** complex forms at equilibrium for domain $a^* < 5$ nt under the reaction condition specified in this work according to NUPACK simulation [27]. In the presence of the split target (**ST**) strand, the local concentrations of **I1** and **I2** increase significantly resulting in the higher probability of them associating momentarily at domain a^* to form a high level of the proper **ST-I1-I2** assembly (major product). The trade-off is that a stable intermediate complex can form between **I1** and **F-Q** at domain b^* (minor product) from which domain a^* now serves as the new reporter trigger toehold. However, due to the short length of domain a^*, the kinetics of strand displacement is slow and equilibrium yield is low.

Conversely, the length of domain b^* can be reduced such it is shorter than domain a^* (Fig. 6.8b). This disrupts the formation of any stable **I1-F-Q** complex. However, due to the short length of domain b^*, strand displacement of the reporter signal is kinetically slow. At the same time, a relatively significant amount of stable **I1-I2** complex is present at equilibrium. This implies that the effect of additional association due to the increased local concentrations of initiator strands in the presence of **ST** is less pronounced.

From the above analysis, we infer that having a longer domain b^* length than domain a^* length is preferred under the constraint that domain a^* should be kept sufficiently short, i.e. <5 nt, such that **I1-I2** complex does not exist at equilibrium. We further supported our hypothetical argument using an experimental example involving two cases where the lengths of domains a^* and b^* were interchanged between 4 and 6 nt (Fig. 6.8c). When the length of domain b^* was longer than domain a^* (as in

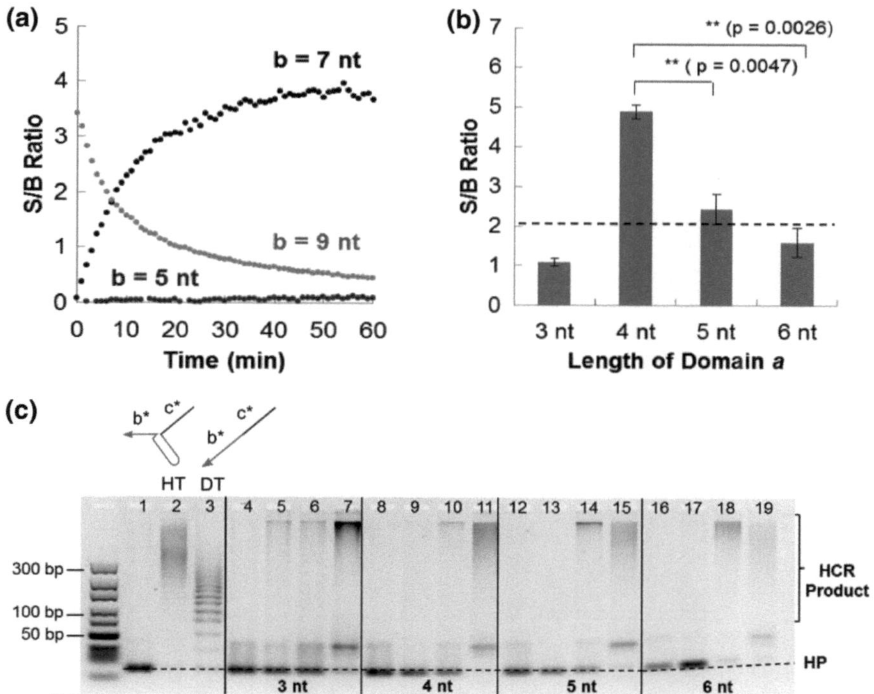

	Controls			Length of domain a*															
				3 nt				4 nt				5 nt				6 nt			
HP1+HP2	+	+	+	+	+	+	+	+	+	+	+	+	+	+	+	+	+	+	+
I1	-	-	-	+	-	+	+	+	-	+	+	+	-	+	+	+	-	+	+
I2	-	-	-	-	+	+	+	-	+	+	+	-	+	+	+	-	+	+	+
T	-	HT	DT	-	-	-	ST	-	-	-	ST	-	-	-	ST	-	-	-	ST

Fig. 6.7 Effect of the length of domain $a*$ and $b*$ on the formation of intermediate complex 1 and 2. **a** The length of domain b* was varied (5 nt, 7 nt and 9 nt) to understand the contribution to the signal-to-background (S/B) ratio by intermediate complex 1. **b** An optimal association length (domain $a*$) of 4 nt resulted in maximum S/B ratio at t = 10 min using the fluorophore-quencher (**F-Q**) readout. The error bars correspond to standard deviation (n = 3). **indicates that $p < 0.005$ for Student's t-test (one-sided). The actual p-values are shown in bracket. **c** Gel electrophoresis was used to visualize the extent of signal and background formation. 500 nM of individual DNA components was used. A 10–300 bp DNA ladder is shown on the left-hand side of the gel

Fig. 6.8a), discernible level of signal was generated over the background. Conversely, when the length of domain $a*$ was longer than domain $b*$ (as in Fig. 6.8b), both signal and background were generated at a similarly low level. These experimental observations were in line with the analysis above. Hence, both the hypothetical argument and experimental results reinforced the importance of keeping the de facto toehold, domain $b*$, at a sufficiently long length (≥6 nt) to ensure good S/B ratio.

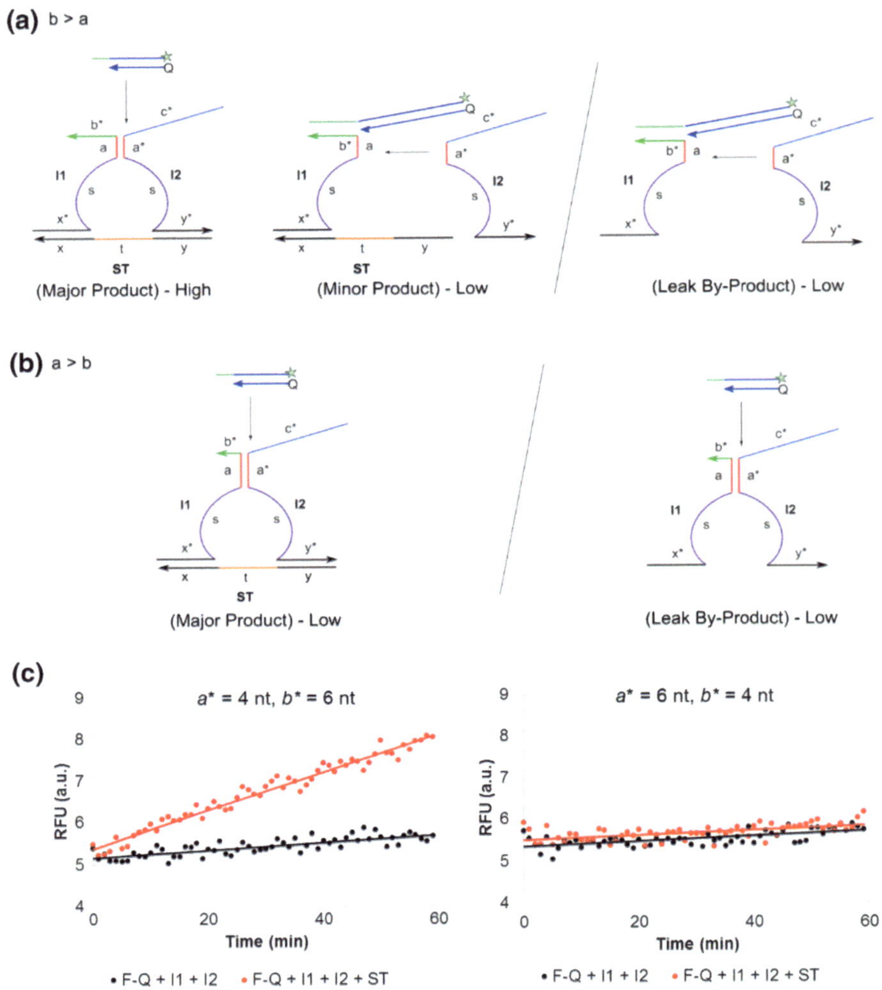

Fig. 6.8 Identification of major and minor products, as well as leak by-products generated when **a** length of domain b^* is greater than domain a^*; **b** length of domain a^* is greater than domain b^*. The terms "high" and "low" indicate the expected level of complex formation relative to the total amount of input DNA strands. The diagonal line partitions the generation of signal (left) and noise (right). **c** The above hypothetical case studies were substantiated using two examples of ($a^* = 4$ nt, $b^* = 6$ nt) and ($a^* = 6$ nt, $b^* = 4$ nt). 100 nM of individual DNA components was used

A better strategy would be to temporally separate the addition of initiator strands (**I1** and **I2**) and reporter strands (**F-Q or HP1** and **HP2**). Under this format, the main design criterion was to reduce the formation of intermediate complex 2 for which the

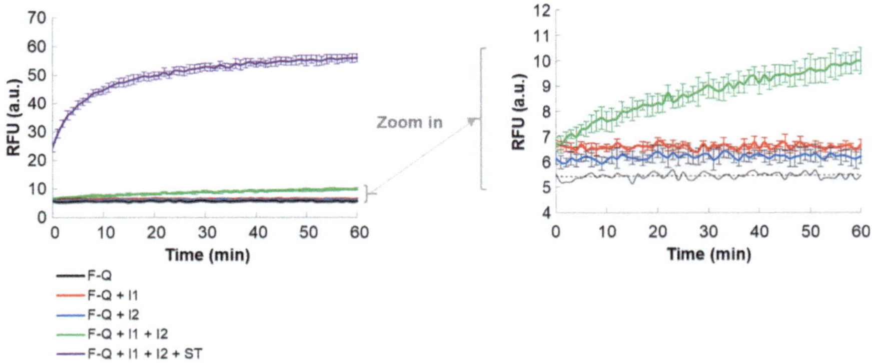

Fig. 6.9 The final DNA split proximity assay was designed with domain $a^* = 4$ nt and domain $b^* = 9$ nt. The generation of positive signal was improved while minimizing the circuit leakage. Unfortunately, Leak II was contributed by an intermediate complex essential for the generation of positive signals, i.e. **I1-F-Q**, and could not be eliminated. The concentration of all DNA components (**F-Q, I1, I2** and **ST**), added whenever relevant, was 100 nM

stability was controlled by the length of domain *a*. 4 nt was found to be the optimal association region length giving the highest S/B ratio (Fig. 6.7b) within as quickly as 10 min without forming appreciable amount of **I1-I2** complex at equilibrium.

Unfortunately, we were unable to eliminate the background noise due to Leak II (Fig. 6.9). This was not surprising as the thermodynamically-favoured **I1-F-Q** complex could form once the reporter strands were introduced. Nonetheless, the involvement of a short domain a^* (4 nt) as the reporter toehold implied exponentially slower reporter triggering as compared to the 9 nt toehold (domain b^*) triggering in the proper **ST-I1-I2** assembly [28]. Overall, we observed that the S/B ratio was not negatively affected even when HCR amplification was involved, except when domain a^* was longer at 6 nt (Fig. 6.7c).

Unfortunately, the exact domain configurations to use require re-optimization under specific experimental conditions since the relative rate of generating signal and background is a kinetics effect. However, the two criteria of having at least a 6 nt length for domain b^* and ensuring that it is longer than domain a^* remain valid under all circumstances that we can think of.

6.2.5 Evaluation of Performance on Model Biomolecular Systems

Having successfully developed a DNA split proximity circuit with minimized circuit leakages, we evaluated its applicability and robustness on a model streptavidin-biotin system. Streptavidin was chosen for its high binding affinity ($K_D \approx 10^{-14}$–10^{-15} nM) to biotin which can be modified onto commercially-available oligonucleotides

(Fig. 6.10a). This allowed us to validate the inherent DNA circuit properties without further complications from the target recognition step which was not the focus of this study.

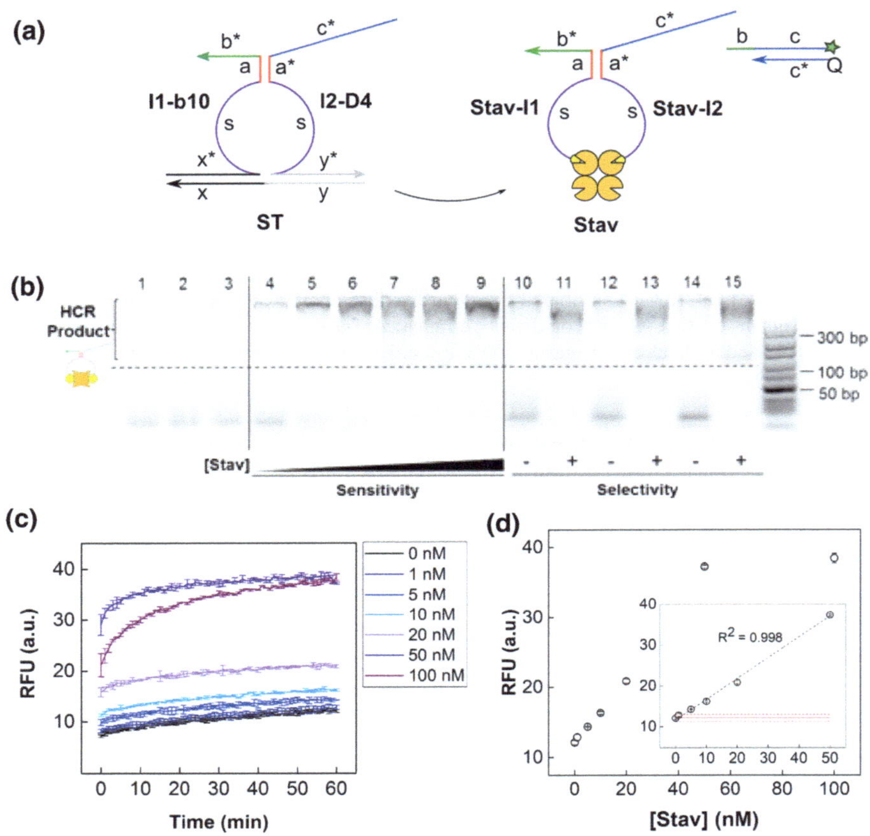

Fig. 6.10 Demonstration of the association toehold system on a streptavidin (Stav)-biotin model. **a** The recognition moiety on the initiators was replaced with biotin for binding with the Stav target. **b** The circuit performance was evaluated using gel electrophoresis and HCR readout. Lane 1 corresponds to hairpins (HP1 and HP2) only, lanes 2 and 3 correspond to controls with HP1 + HP2 + I1 and HP1 + HP2 + I2 respectively. Lanes 4–9 represent the signal developed in presence of 0 nM, 10 nM, 50 nM, 100 nM, 250 nM and 500 nM of Stav. Lanes 10–15 represent the circuit selectivity in presence of interference from 500 nM BSA (lanes 10 and 11), 5000 nM BSA (lanes 12 and 13) and 10% FBS (lanes 14 and 15). 500 nM of individual DNA circuit components was used. A 10–300 bp DNA ladder is shown on the right-hand side of the gel. **c** Signal trace over time when varying concentration of Stav (0–100 nM) was added to the association toehold system. Error bars correspond to standard deviation (n = 3). **d** The equilibrium RFU signal at t = 1 h was plotted for the range of Stav concentrations tested. A linear trend was observed between 0–50 nM (inset). The solid red line corresponds to the mean background noise while the dotted red line corresponds to 3 standard deviations from the mean noise level

The HCR product bands were visibly more intense than the negative control (lane 4) for the range of streptavidin concentrations tested (10–500 nM, lanes 5–9) (Fig. 6.10b). The circuit remained selective under the interferences by non-specific bovine serum albumin (BSA) protein (lanes 10–13) and a biological environment of 10% fetal bovine serum (FBS) (lanes 14 and 15). The detection result was further quantified using **F-Q** readout signal (Fig. 6.10c). Clear dosage dependence was observed for 0–50 nM streptavidin target with a limit of detection (LOD) at 2.95 nM (Fig. 6.10d). These results demonstrated the good sensitivity, selectivity, compatibility with biological environment and reasonable detection speed (10 min) of the association toehold system.

6.3 Conclusion

In conclusion, a simple association toehold system for signal transduction was engineered through a systematic approach aimed primarily at minimizing circuit leakages. The key concepts introduced in this work, i.e. inter-domain bridging and 3′-truncation as the main source of defect in solid-state oligonucleotide synthesis, are applicable to any DNA circuits in general. Particularly, the inter-domain bridging concept was effective in improving both the migration kinetics and equilibrium signal while suppressing asymptotic leakage, which was essentially a triple-win design strategy.

The final design with minimized circuit leakage was applied on a model streptavidin-biotin system which returned decent performance in terms of sensitivity, selectivity and robustness even in a biological environment. We acknowledge that the streptavidin-biotin model presented in this chapter has a stronger binding affinity than most biomolecular interactions. We will further demonstrate the use of our overall circuit for the interrogation of a wider breadth of biological proximity events in the next chapter.

The crux of using the association toehold as a general signal transduction module lies in the sequence design being independent of the specific target(s). Therefore, the sequences evolved from this study can be used directly as initiator probes upon conjugation to desired recognition moieties, e.g. antibody, peptide and aptamers, to probe for diverse biomolecular interactions. The association toehold concept is a simple yet robust signal transduction strategy which can potentially be interfaced with other downstream modules to perform more complex operations, e.g. releasing DNA-bound drug molecule.

References

1. Zhou, W., Jimmy Huang, P.-J., Ding, J., Liu, J.: Aptamer-based biosensors for biomedical diagnostics. Analyst **139**, 2627 (2014)
2. Morell, M., Ventura, S., Avilés, F.X.: Protein complementation assays: approaches for the in vivo analysis of protein interactions. FEBS Lett. **583**, 1684 (2009)
3. Stynen, B., Tournu, H., Tavernier, J., Van Dijck, P.: Diversity in genetic in Vivo methods for protein-protein interaction studies: from the yeast two-hybrid system to the mammalian split-luciferase system. Microbiol. Mol. Biol. R. **76**, 331 (2012)
4. Porter, J.R., Stains, C.I., Jester, B.W., Ghosh, I.: A general and rapid cell-free approach for the interrogation of protein—protein, protein—DNA, and protein—RNA interactions and their antagonists utilizing split-protein reporters. J. Am. Chem. Soc. **130**, 6488 (2008)
5. Genot, A.J., Zhang, D.Y., Bath, J., Turberfield, A.J.: Remote toehold: a mechanism for flexible control of DNA hybridization kinetics. J. Am. Chem. Soc. **133**, 2177 (2011)
6. Chen, X.: Expanding the rule set of DNA circuitry with associative toehold activation. J. Am. Chem. Soc. **134**, 263 (2012)
7. Tang, Y., Lin, Y., Yang, X., Wang, Z., Le, X.C., Li, F.: Universal Strategy to Engineer Catalytic DNA Hairpin Assemblies for Protein Analysis. Anal. Chem. **87**, 8063 (2015)
8. Li, F., Lin, Y., Le, X.C.: Binding-induced formation of DNA three-way junctions and its application to protein detection and DNA strand displacement. Anal. Chem. **85**, 10835 (2013)
9. Tang, Y., Wang, Z., Yang, X., Chen, J., Liu, L., Zhao, W., Le, X.C., Li, F.: Constructing real-time, wash-free, and reiterative sensors for cell surface proteins using binding-induced dynamic DNA assembly. Chem. Sci. **6**, 5729 (2015)
10. Zuker, M.: Mfold web server for nucleic acid folding and hybridization prediction. Nucleic Acids Res. **31**, 3406 (2003)
11. Li, B., Ellington, A.D., Chen, X.: Rational, modular adaptation of enzyme-free DNA circuits to multiple detection methods. Nucleic Acids Res. **39**, e110 (2011)
12. Machinek, R. R. F.; Ouldridge, T. E.; Haley, N. E. C.; Bath, J.; Turberfield, A. J. Programmable Energy Landscapes for Kinetic Control of DNA Strand Displacement. *Nat. Commun.* **5** (2014)
13. Zhang, D.Y., Winfree, E.: Robustness and modularity properties of a non-covalent DNA catalytic reaction. Nucleic Acids Res. **38**, 4182 (2010)
14. Chen, X., Briggs, N., McLain, J. R., Ellington, A. D.: Stacking nonenzymatic circuits for high signal gain. Proc. Natl. Acad. Sci. U.S.A. **110**, 5386 (2013)
15. Ang, Y.S., Yung, L.-Y.L.: Rational design of hybridization chain reaction monomers for robust signal amplification. Chem. Commun. **52**, 4219 (2016)
16. Dirks, R.M., Pierce, N.A.: Triggered amplification by hybridization chain reaction. Proc. Natl. Acad. Sci. U.S.A. **101**, 15275 (2004)
17. Temsamani, J., Kubert, M., Agrawal, S.: Sequence identity of the n-1 product of a synthetic oligonucleotide. Nucleic Acids Res. **23**, 1841 (1995)
18. Li, B., Jiang, Y., Chen, X., Ellington, A.D.: Probing spatial organization of DNA strands using enzyme-free hairpin assembly circuits. J. Am. Chem. Soc. **134**, 13918 (2012)
19. Jiang, Y.S., Bhadra, S., Li, B., Ellington, A.D.: Mismatches improve the performance of strand-displacement nucleic acid circuits. Angew. Chem. Int. Ed. **53**, 1845 (2014)
20. Schofield, M.J., Hsieh, P.: DNA mismatch repair: molecular mechanisms and biological function. Annu. Rev. Microbiol. **57**, 579 (2003)
21. Hughes, R.A., Miklos, A.E., Ellington, A.D.: In Christopher, V (ed.) Methods Enzymol, vol. 498, p. 277. Academic Press (2011)
22. Ducani, C., Kaul, C., Moche, M., Shih, W.M., Hogberg, B.: Enzymatic production of 'monoclonal stoichiometric' single-stranded DNA oligonucleotides. Nat. Methods **10**, 647 (2013)
23. Khodakov, D.A., Khodakova, A.S., Huang, D.M., Linacre, A., Ellis, A.V.: Protected DNA strand displacement for enhanced single nucleotide discrimination in double-stranded DNA. Sci. Rep. **5**, 8721 (2015)

24. Srinivas, N., Ouldridge, T.E., Šulc, P., Schaeffer, J.M., Yurke, B., Louis, A.A., Doye, J.P.K., Winfree, E.: On the biophysics and kinetics of toehold-mediated dna strand displacement. Nucleic Acids Res. **41**, 8886 (2013)
25. Davis, Mark E., Davis, R.J.: Fundamentals of chemical reaction engineering. McGraw-Hill, Boston (2003)
26. Zhang, D.Y., Seelig, G.: Dynamic DNA nanotechnology using strand-displacement reactions. Nat. Chem. **3**, 103 (2011)
27. Zadeh, J.N., Steenberg, C.D., Bois, J.S., Wolfe, B.R., Pierce, M.B., Khan, A.R., Dirks, R.M., Pierce, N.A.: NUPACK: analysis and design of nucleic acid systems. J. Comput. Chem. **32**, 170 (2011)
28. Zhang, D.Y., Winfree, E.: Control of DNA strand displacement kinetics using toehold exchange. J. Am. Chem. Soc. **131**, 17303 (2009)

Chapter 7
DNA Split Proximity Circuit as a General Platform for Interrogating Biomolecular Events

7.1 Introduction

The design aspects of the individual circuit modules have been conceptualized, validated and discussed in the previous chapters. The aim of this chapter is to integrate the different modules into an overall circuit for detecting biomolecules and biomolecular events while preserving the self-contained and one-pot feature from the prototype introduced in Chap. 4, using the reassembly of the split domains as the mechanism for circuit activation. We name this the "split proximity circuit" (Scheme 7.1).

The split proximity circuit is modular and general in its operation. The reassembled trigger strand (c^* b^*) upon target activation can be linked to any downstream module that operates via toehold mediated strand displacement. For the scope of this chapter, we are interested in detecting different types of biomolecules simply by attaching the corresponding recognition moiety to the target(s) of interest. Two types of readout system were used as the downstream module, i.e. HCR FRET and fluorophore-quencher (**F-Q**).

In this chapter, we first optimized the operating conditions for our integrated split proximity circuit (Sects. 7.2.1 and 7.2.2). The kinetics of the individual steps involved was characterized and the reassembly of the split domains was identified as the rate limiting step (Sect. 7.2.3). Finally, the split proximity circuit was used to detect a range of biomolecules by plugging in the relevant recognition moiety, *i.e.* complementary DNA to detect single nucleotide mutation in let-7 microRNA (miRNA) family (Sect. 7.2.4), aptamers to characterise thrombin binding (Sects. 7.2.5 and 7.2.6), and antibodies to detect receptor protein dimers (Sect. 7.2.7).

We demonstrated through these studies that the split proximity circuit can be used for varied purposes, from simple end-point detection to semi-quantitative analysis of the binding affinity and probing of interaction events. Notably, the circuit operated robustly despite the varying reaction conditions, e.g. inclusion of K^+ for thrombin detection and in unpurified cell lysates.

A. Yan Shan, *Engineering a Robust DNA Circuit for the Direct Detection of Biomolecular Interactions*, Springer Theses,
https://doi.org/10.1007/978-981-13-2188-7_7

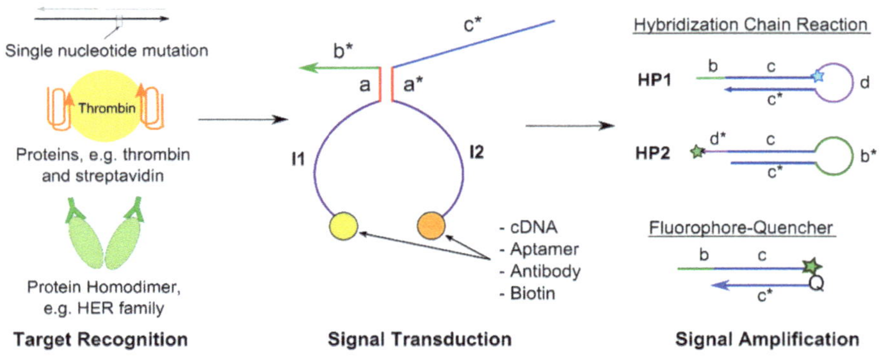

Scheme 7.1 The split proximity circuit can be used as a general platform for detecting different types of target for various applications by changing the recognition moiety and readout module in a "plug-and-play" format

Table 7.1 List of domain sequences in the current split proximity circuit design

Domain	Description	Length (nt)	Sequence
a	Association region	4	GTG C
b	Reporter toehold	9	GTT GGA ATT
c	Readout branch migration site	12	GGG AGT AAG GGC
d	Alternating toehold of HCR	6	TGT GAT
s	Spacer	15 (flexible)	TTT TTT TTT TTT TTT

7.2 Results and Discussion

7.2.1 Design Framework for Generating Circuit Sequences

The current split proximity circuit is ready for use in a "plug-and-play" format and the respective domains sequences are shown in Table 7.1. If necessary, the general domain specifications can be used to re-design new sequences using NUPACK simulation based on the design guidelines below (Scheme 7.2).

One should first decide if fast kinetics is crucial for the operation, at the expense of increased risk of circuit leakage. If speed is crucial, a shorter branch migration length (domain c^*) and longer toehold (domain b^*) can be used while keeping the toehold length within 12 nt. If HCR is used as the readout module, the C·G content of the toehold (domain b^* and d) should be restricted to below 30–40% (Sect. 5.2.2). The stem length (domain c^*) must be longer than the toehold, and is recommended to have a C·G content of at least 60%. The HCR hairpins (**HP1** and **HP2**) are now fully defined.

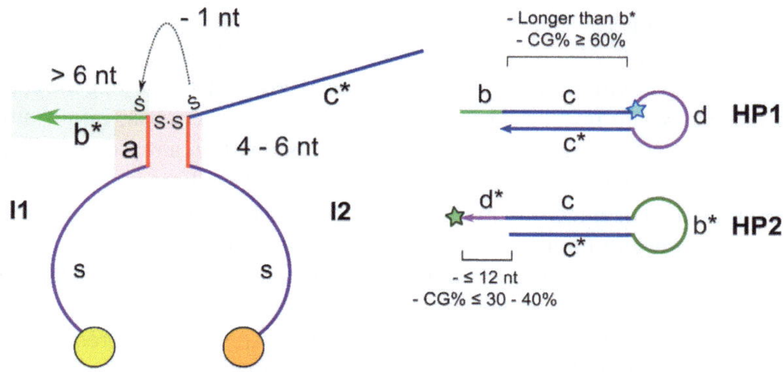

Scheme 7.2 Summary of key design guidelines to generate robust sequences for the split proximity circuit

For the signal transduction module design, the first nucleotide on domain c^* at the three-way junction is shifted to the adjacent position on domain b^* to minimize undesired strand displacement by initiator 2 (**I2**). A two-thymine bulge is included between domain b^* and a to stabilize the three-way junction and promote strand displacement kinetics [1]. Preferably, G·C base pairs are used around the three-way junction for additional stabilization. The association region (domain a) should be kept short such that its melting temperature in absence of target is less than 10 °C. As a rough guide, this translates to 4–6 nt length for approximately 50% C·G content, and at most 8 nt for pure A·T composition; though it should strictly be kept shorter than domain b^* (Sect. 6.2.4).

An example of the sample NUPACK code is presented in Appendix B. 10 sets of random sequence designs were generated from the simulation within the time scale of seconds. The sequences were arranged in order of increasing ensemble defect and screened manually for the above considerations. The set(s) of sequences fitting all criteria were shortlisted and analyzed on NUPACK to confirm the lack of background complexes, e.g. **HP1-HP2-HP1-...** for a mixture of **HP1** and **HP2**, and **I1-I2-HP1** for a mixture of **I1**, **I2**, **HP1** and **HP2**. The set of sequences which generated the highest equilibrium concentration of the desired product(s) was selected for experimental testing.

7.2.2 Optimizing Circuit to Operate Under Physiological Conditions

DNA circuit typically requires high salt conditions to perform its operations (Sect. 2.5.3). We previously challenged the individual modules under high salt condition, e.g. 750 mM NaCl for HCR, and 480 mM NaCl and 5 mM $MgCl_2$ for association toehold, to test their robustness against leakages. For actual application works, an environment closer to physiological condition is preferred since most biomolecules are sensitive to its environment [2, 3]. This section deals with developing the upstream circuit design, which underwent more stringent selection criteria, into a practical assay where the generation of signal over background within a reasonable time was of greater priority than minimizing the absolute level of background.

The parameters deemed important for consideration included environmental factors, e.g. salt concentration, concentration of circuit components and temperature, as well as the association region length. To be concise, key trends were highlighted and the detailed optimization process was left out.

1X PBS (pH 7.4) was used as the basic buffer system. The current split proximity circuit consisted of a 4 nt association length which was the minimum length required for effective signal transduction (Sect. 6.2.4). However, the generation of positive signal was slow under this low salt condition (Fig. 7.1), and is not suitable for applications requiring stringent environment such as live cell assays. Magnesium is a more effective cation than sodium [4], and a gradual increase in the Mg^{2+} concentration resulted in a drastic improvement to the signal-to-background (S/B) ratio. The best S/B ratio was obtained for 5 mM Mg^{2+} while the kinetics was improved by approximately 7-fold at 10 mM Mg^{2+}.

We have shown in Chap. 5 that HCR was not particularly sensitive to Na^+ and Mg^{2+} concentrations. It is likely that the main role of Mg^{2+} was to stabilize the three-way junction during the reassembly of split domains. For operations requiring low salt condition, the association length can be extended to the upper limit of 6 nt. The issue of leakage was less problematic under low salt and circuit component concentrations and high S/B ratio was achieved except at the highest Mg^{2+} concentration of 10 mM.

Next, we studied the effect of varying the reaction ratios between the hairpins and initiators, as well as between **HP1** and **HP2**. We set 20 nM of **I1** and **I2** as the basis of comparison. The **HP1**: **HP2** ratio of 2:1 generated the highest FRET signal (Fig. 7.2a–c), in line with other reports where HCR is promoted by having a higher molar ratio of **HP1** to **HP2** [5, 6]. This is due to the unequal participation of **HP1** and **HP2** in HCR, where **HP1** is responsible for both initiating HCR and propagating the growth of the assembly product (Appendix F). Having more **HP1** input ensured that it was not exhausted prematurely, especially in presence of high trigger concentration. The difference narrowed with decreasing trigger concentration as the proportion of **HP1** bound to the trigger strand became negligible relative to the total input concentration. Adding more HCR hairpins relative to the initiators did not improve the signal (Fig. 7.2d). For the rest of the thesis, the general reaction condition of 40 nM **HP1**, and 20 nM **HP2, I1** and **I2** in 1X PBS (pH 7.4) and 10 mM $MgCl_2$ was used, unless stated otherwise.

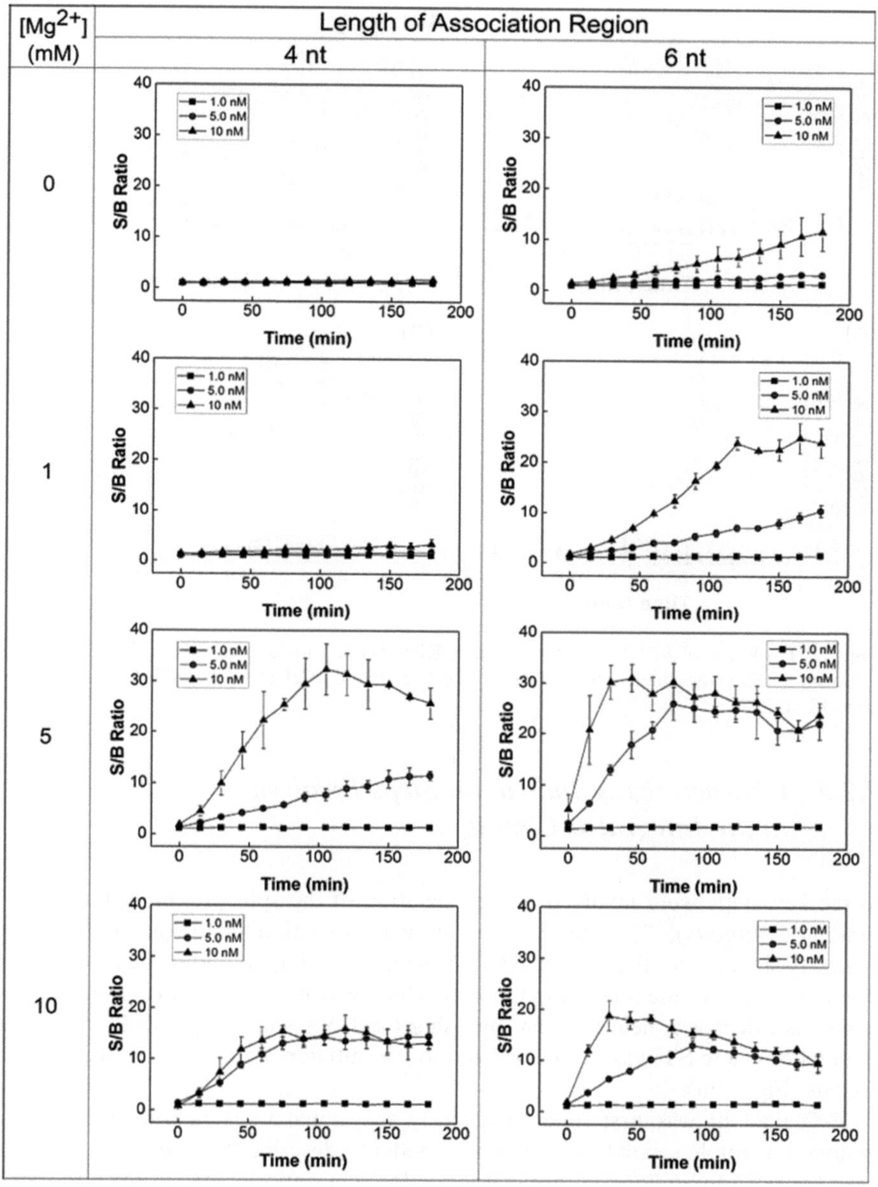

Fig. 7.1 Kinetics of the split proximity circuit evaluated under different Mg^{2+} concentrations for two association lengths of 4 and 6 nt at 25 °C. The circuit performance was expressed in terms of its signal-to-background (S/B). Signal is defined as the FRET ratio obtained in presence of 1.0, 5.0 and 10 nM of split trigger (**ST**); while background is the FRET ratio obtained in absence of **ST**. All data are shown as mean ± S.D (n = 3)

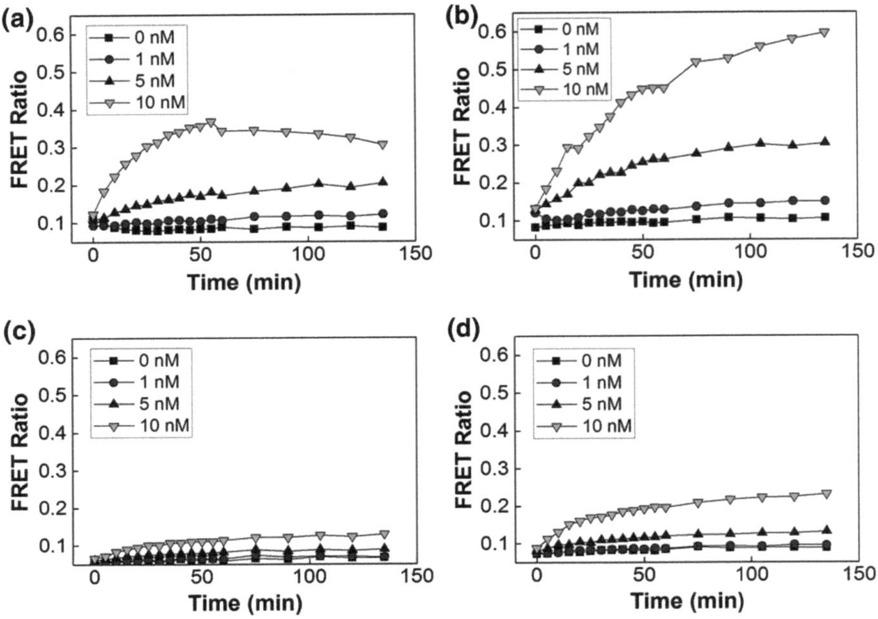

Fig. 7.2 Evolution of FRET ratio for different **HP1: HP2: I** molar ratios. 0, 1, 5 and 10 nM of split trigger (**ST**) was added to **a** 1:1:1, **b** 2:1:1, **c** 1:2:1 and **d** 2:2:1 of **HP1: HP2: I** on the basis of 20 nM **I** at 25 °C

7.2.3 Characterizing Individual Steps Involved in the Integrated Circuit

Three key steps were involved in the operation of the split proximity circuit: (1) binding to target(s), (2) signal transduction via association toehold and (3) readout, typically signal amplification via HCR (Scheme 7.1). Target binding depends on the binding affinity of the recognition moiety which was irrelevant to the circuit design itself. Its effect was held constant by using the DNA split trigger (**ST**) system as the model where **ST** was pre-hybridized to the initiators before adding the readout signals (Fig. 7.3a).

F-Q readout was first used to quantify the signal transduction kinetics. We assumed a simple second order reaction model to describe the system as typically used for DNA hybridization. A much lower **DT** concentration of 2.0 nM was added to a mixture of 40 nM **HP1** and 20 nM **HP2** which effectively reduced the system to a pseudo first order reaction. The pseudo first order rate constant (k') was estimated through curve fitting from which the actual second order rate constant (k) was calculated. A 9 nt toehold (domain b^*) in the direct trigger (**DT**) was sufficient to drive strand displacement to completion, where approximately 2.0 nM of fluorophore was released from 2.0 nM of **DT** inputted (Fig. 7.3b). The measured hybridization rate

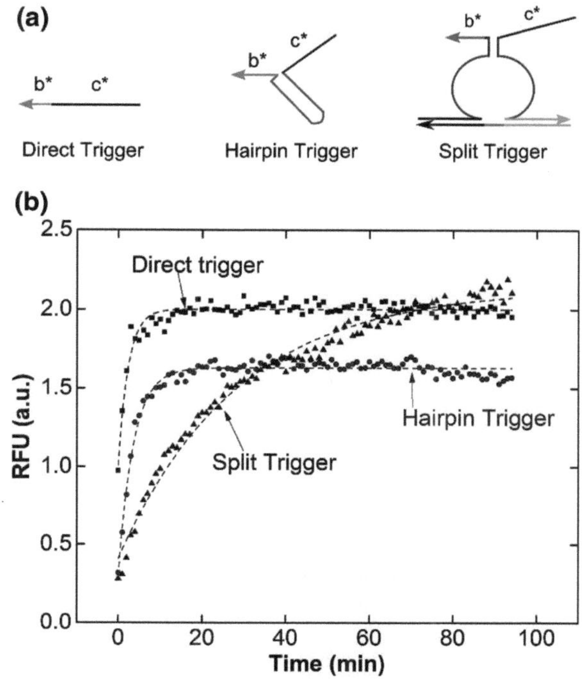

Fig. 7.3 Effect of the split in trigger domains on the hybridization kinetics. **a** Schematics of the three types of trigger strand used where the domains became increasingly separated. **b** Circuit kinetics when 2.0 nM of direct trigger, hairpin trigger and split trigger was added respectively to 20 nM of **F-Q** complex

Table 7.2 Kinetics parameters obtained when the RFU was fitted to a pseudo first order reaction kinetics for three systems of direct trigger, hairpin trigger and split trigger

Parameters	System		
	Direct trigger	Hairpin trigger	Split trigger
C_{eq} (nM)	2.00	1.63	2.14
k' (min^{-1})	0.472	0.277	0.0353
k (10^5 M^{-1} s^{-1})	3.62	2.32	0.293

C_{eq} refers to the signal concentration at equilibrium
k' is the pseudo first order rate constant and k is the calculated second order rate constant

was of similar order of magnitude (10^5 M^{-1} s^{-1}) as that reported in literature for solution phase DNA hybridization (Table 7.2) [7].

The effects of splitting the domain was studied by separating domains $b*$ and $c*$ on **DT** into two segments held together as a hairpin structure, or hairpin trigger (**HT**) (Fig. 7.3a). The overall signal profile generated was similar to **DT** albeit with a 25% reduction in the equilibrium signal (Fig. 7.3b). This was due to the formation of undesired secondary structure in **HT** which blocked the active domains (domains $b*$ and $c*$) from undergoing strand displacement (Fig. 7.4). However, NUPACK simulation accounted only for the most probable secondary structure which formed with 91.6% probability in this case.

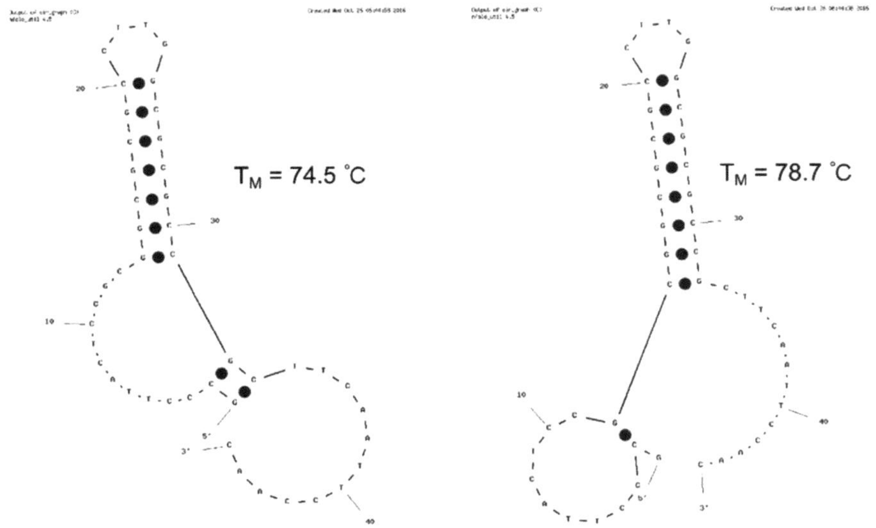

Fig. 7.4 Formation of undesired secondary structures of the hairpin trigger which blocked the active domains *b** and *c**. The structures and their corresponding melting temperatures (T_M) were obtained from m fold simulation

Table 7.3 Kinetics parameters obtained when the RFU was fitted to a pseudo first order reaction kinetics for the HCR and signal transduction steps, and overall split proximity circuit operation

Parameters	System		
	DT-HCR	ST-F-Q	ST-HCR
k' (min^{-1})	0.117	0.0353	0.0257
k (10^4 M^{-1} s^{-1})	9.75	2.93	2.14

The separation of domains *b** and *c** on different initiator strands in the **ST** system greatly penalized the strand displacement kinetics by approximately 12-fold while the signal eventually developed towards the equilibrium value attained by **DT** (Table 7.2). The half-life of the signal transduction reaction ($\tau_{reaction}$) was approximately 20 min which was a reasonable time scale for detecting biomolecular interactions which conventionally involve time-consuming procedures.

Next, **DT** was added to **HP1** and **HP2** to characterize the HCR kinetics; **ST** added to the split proximity circuit to characterize the overall circuit kinetics. Signal transduction was approximately 3.5-fold slower than HCR, and was therefore the rate limiting step (Table 7.3). Indeed, the kinetics of the overall circuit followed that of signal transduction closely (Fig. 7.5). The overall reaction rate of our split proximity circuit, without considering the target binding kinetics, was 2.14×10^4 M^{-1} s^{-1} with an approximate reaction half-life of 27 min.

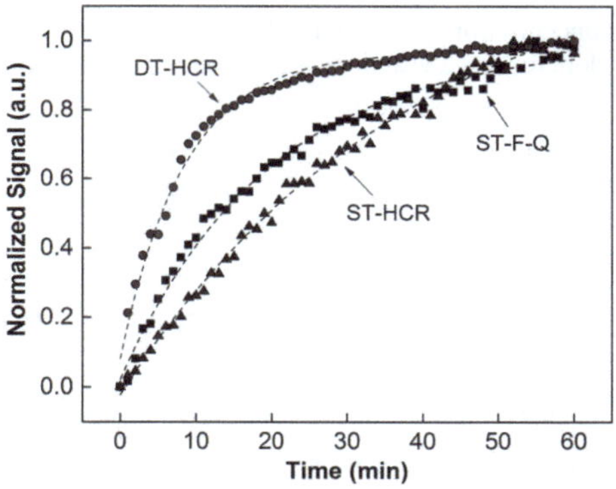

Fig. 7.5 Characterisation of the individual steps involved in the split proximity circuit. The HCR and **F-Q** signals were normalized to facilitate comparison. 2 nM of the indicated trigger (**DT** or **ST**) was reacted with the relevant readout signal

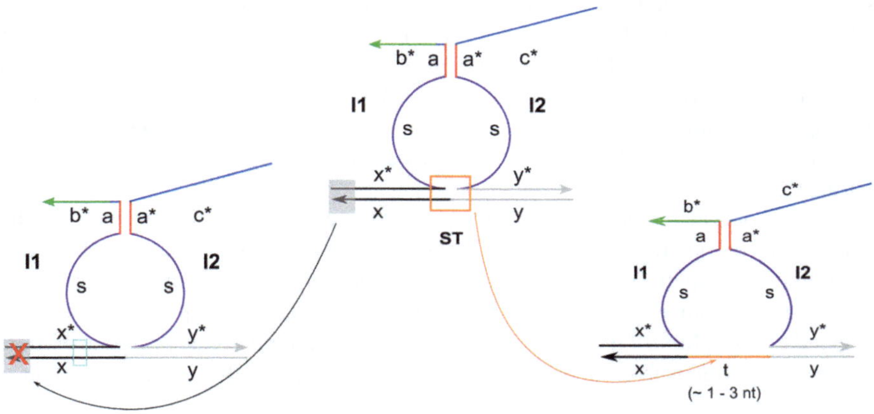

Scheme 7.3 The split at the recognition site can be further engineered to discriminate nucleotide mutations (blue box) by changing the recognition length (red cross) (left) or to improve the readout signal by relaxing the spatial constraints (right)

7.2.4 *Detection of Single Nucleotide Mutation*

An obvious extension to the **ST** system was for nucleic acids detection, particularly for single nucleotide mutations. The lower half of the split proximity circuit containing the recognition moiety (domain x^* and y^*) could be further engineered based on two design considerations, i.e. the degree of spatial constraint at the junction of initiators binding and the effect of the recognition length on duplex stability (Scheme 7.3). Each permutation of **ST** is denoted as **ST** "length of domain y"-"length of spacer t"-"length of domain x" (from 5′- to 3′-end) in the following discussion.

Table 7.4 Equilibrium concentration of **ST-I1-I2** obtained for different split trigger (**ST**) designs with different lengths of domain *x* and *y*, and increasing number of thymine(s) between them

	Domain *t*	Split trigger design				
		ST8-t-8	ST9-t-9	ST10-t-10	ST11-t-11	ST12-t-12
[ST-I1-I2] (nM)	0 T	0	0.25	5.50	12.0	14.0
	+ 1T	0.25	3.90	15.0	18.0	**19.0**
	+ 2T	0.86	7.60	**17.0**	**19.0**	19.0
	+ 3T	**1.10**	**8.40**	17.0	19.0	19.0
	+ 4T	0.88	7.60	17.0	19.0	19.0

Results were obtained from NUPACK simulation
The highest [**ST-I1-I2**] for each design was bolded

Domains *x** and *y** on the original initiator strands were fully complementary to **ST**, creating a crowded environment about the junction where the three DNA strands (**I1**, **I2** and **ST**) met. The lack of conformational freedom hindered base-stacking and destabilized the **ST-I1-I2** complex [8]. The spatial constraint was relaxed by introducing a short stretch of thymine nucleotides between the two recognition sites. Adding one thymine increased the equilibrium concentration of **ST-I1-I2** complex from 14 to 19 nM (Table 7.4). Further increase in the number of thymine was of marginal effect.

To discriminate single nucleotide mutation, the recognition domains were gradually reduced to a critical hybridization length where further omission of one base pair disrupted the formation of **ST-I1-I2** complex. The critical length for the current **ST** sequence was from 9 to 8 nt, where the equilibrium **ST-I1-I2** complex concentration dropped to zero (Table 7.4). Also, a less stable complex required the presence of more thymine nucleotides to counteract the destabilizing effect of the spatial constraint at the junction.

The simulation results were validated experimentally using selected sequences (Fig. 7.6). No FRET signal developed for short **ST** domain lengths of 8 nt (**ST8-0-8**), which increased appreciably for longer lengths of 9 nt (**ST9-0-9**) and 12 nt (**ST12-0-12**). The inclusion of a single thymine between the two recognition sites (**ST12-1-12**) further boosted the FRET signal.

Next, the positional effect of the mismatch on the discrimination ability of the circuit was investigated, especially for end mutations which are challenging since they disrupt the duplex stability to the least extent [9]. We reduced the length of domain *x* to the critical length of 9 nt and introduced three separate single nucleotide mismatches (SM) at different positions along domain *x* (Scheme 7.4). NUPACK was used to simulate the various scenarios.

SM2, with its midpoint mutation, was most disruptive to the formation of **ST-I1-I2** (Table 7.5). End mismatch (**SM1**) was comparable to removing one nucleotide from **ST**, or **ST12-0-8**. This design was not good for discriminating mismatches at

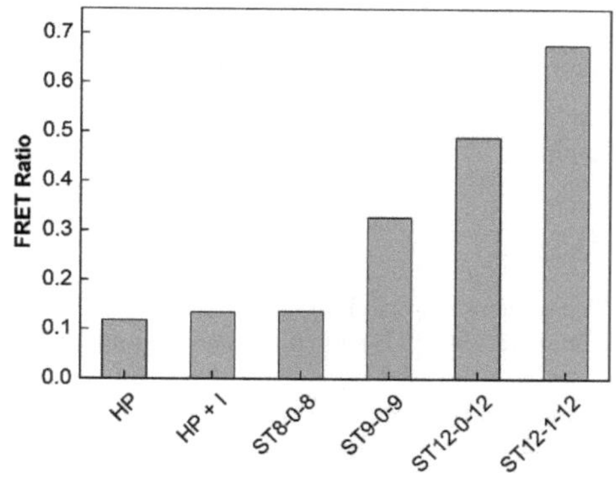

Fig. 7.6 FRET ratio obtained when 20 nM of the listed **ST** designs were reacted with the split proximity circuit

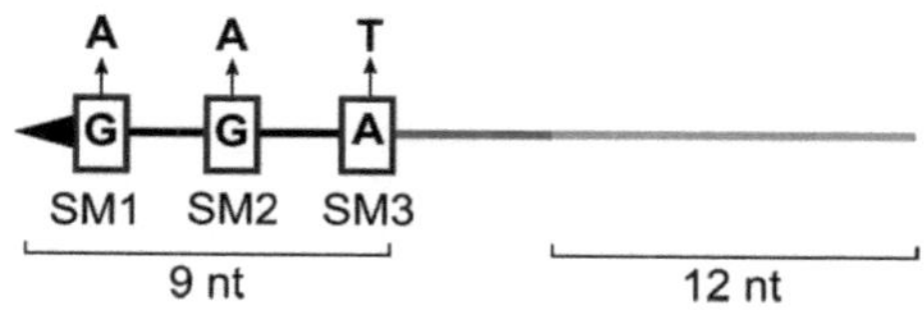

Scheme 7.4 A synthetic target comprising of single nucleotide mismatches (SM) at different positions of the shorter domain *x*

Table 7.5 Equilibrium concentration of **ST-I1-I2** obtained for different synthetic target sequences of various lengths and single mismatch (SM) positions

	Domain *t*	Split trigger design				
		ST12-t-9	ST12-t-8	SM1	SM2	SM3
[ST-I1-I2] (nM)	0 T	0.41	0	0	0	0.39
	+ 1T	4.80	0.33	0.34	3.60×10^{-2}	5.20
	+ 2T	8.9	1.10	1.10	0	4.70
	+ 3T	10.0	1.50	1.50	0	4.50
	+ 4T	9.0	1.10	1.10	0	4.30

Results were obtained from NUPACK simulation

the junction where the three strands bind (**SM3**) which instead relaxed the spatial constraint at the hybridization junction.

The split proximity circuit was then used for detecting microRNAs (miRNAs) which are short, non-coding RNA sequences (typically 22 nt in length) responsible for regulating the post transcriptional expression of messenger RNA (mRNA) [10]. Conventional detection methods, such as quantitative real-time polymerase chain reaction (qRT-PCR), is challenging due to its short length and close similarity between members of the same family [11]. The let-7 miRNA family, which members differed by one or two nucleotides, was used as a proof-of-concept model (Table 7.6) [12].

We modified domains x^* and y^* of the original **I1** and **I2** to discriminate let-7a miRNA against other let-7 miRNA family members. The target sequence in an

Table 7.6 List of microRNA (miRNA) sequences of the let-7 family

miRNA	Sequence
let7a	UGA GGU AGU AGG UUG UAU AGU U
let7b	UGA GGU AGU AGG UUG U**<u>G</u>**U **<u>G</u>**GU U
let7c	UGA GGU AGU AGG UUG UAU **<u>G</u>**GU U
let7d	**<u>A</u>**GA GGU AGU AGG UUG **<u>C</u>**AU AGU U

The sequences are colour coded according to Scheme 7.4
The mutation site is shown in bold and underlined

Table 7.7 Configuration of domain design for discriminating let-7a miRNA against other let-7 miRNA family members and the corresponding equilibrium **ST-I1-I2** concentration obtained from NUPACK simulation. The domains are shown in Scheme 7.3

Domain (nt)			[ST-I1-I2] (nM)			
x^*	y^*	t	let7a	let7b	let7c	let7d
11	11	0	5.0	0	0.2	0
10	11	1	0.038	0	0	0
11	10	1	8.0	0	0.37	2.9×10^{-2}
10	10	2	0.22	0	0	0

actual biological system is fixed and so the relaxation of spatial constraint at the junction was executed by excluding nucleotide(s) from the recognition domains on the initiators.

Most of the mutation sites were located on domain x^* with a lower G·C content and a critical length of 11–10 nt (Table 7.7). The domain design of $x^* = 11$ nt, $y^* = 10$ nt, $t = 1$ nt was selected since it generated the highest concentration of **ST-I1-I2** complex. Slightly higher discrimination ratio, *i.e.* ratio of signal from let-7a to let-7c, could be achieved using the design without any spatial relaxation. However, the effective **ST-I1-I2** concentration from the reaction with let-7c and let-7d (highest being 0.2 nM) was close to the LOD of HCR FRET readout (Sect. 5.2.5), i.e. one-hundredth of hairpin concentration where [**HP1**] = 40 nM and [**HP2**] = 20 nM, and might not result in appreciable background signal.

The initiators design was tested experimentally. The FRET signal from let-7a miRNA was significantly higher ($p < 0.005$) than the negative control (no target) after 1 h of reaction (Fig. 7.7a). No significant FRET signal was generated from other let-7 family members, even for the most problematic let-7c target. The split proximity circuit picked up significant signal from let-7a miRNA target in mixtures containing other family members. However, the FRET signal generated was significantly lower ($p < 0.05$) than when pure let-7a miRNA was used. This could be an issue for practical applications especially if quantification were required.

The limit of detection (LOD) was 0.714 nM of let-7a (Fig. 7.7b). This sub nanomolar LOD was at the upper limit of reported solution-phase methods which can be as low as femtomolar [13–15] range. The key advantage of our split proximity circuit

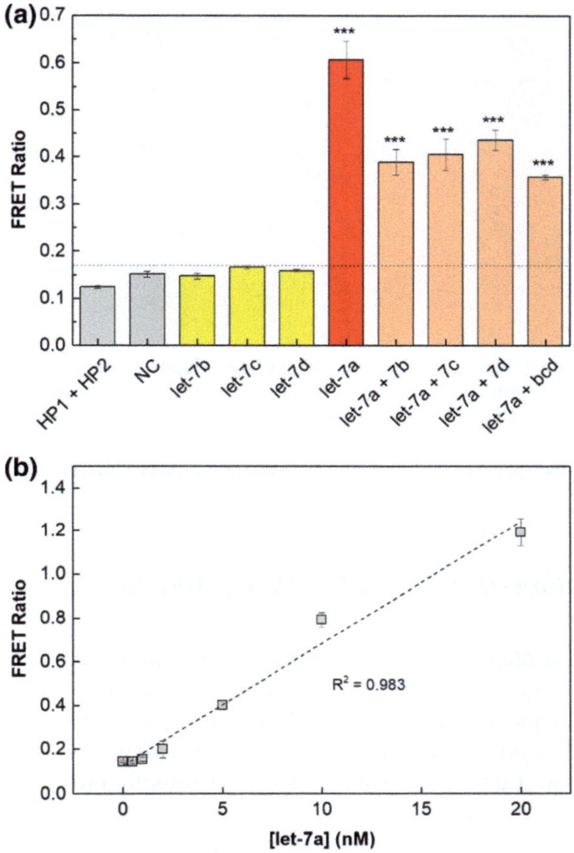

Fig. 7.7 Discriminating the let-7 miRNA family using the split proximity circuit in. **a** Selectivity of the circuit was determined by mixing 5 nM of the respective let-7 members. FRET signal was measured after 1 h of circuit operation. ***refers to $p < 0.005$ (n = 3). The dotted black line indicates the circuit limit of detection, defined as three times S.D. above the mean of negative control (NC). The bar chart is colour coded as: grey—controls, yellow—single mismatched sequences, red—perfectly matched target, and orange—mixture of sequences. **b** Dosage relationship between the FRET ratio and let-7a miRNA concentration. All data are shown as mean ± S.D (n = 3)

is its simple, one-pot format. A comparable platform based on bioluminescence resonance energy transfer (FRET) between Rluc and quantum dot had a similar LOD range of 20 nM [16].

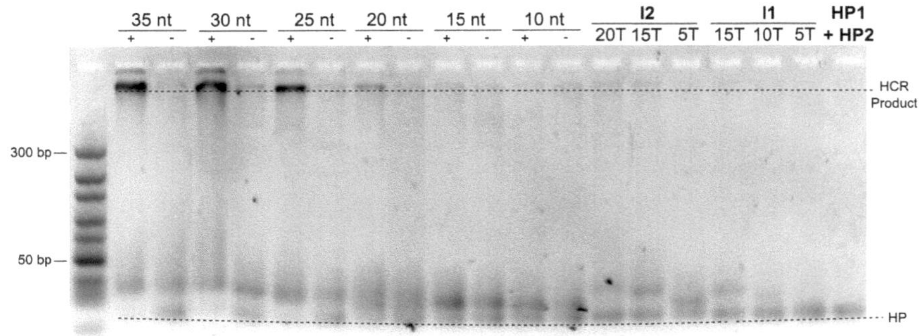

Fig. 7.8 Gel electrophoresis analysis of the effect of the initiator spacer length on thrombin detection. 500 nM of α-thrombin was reacted with 500 nM of the split proximity circuit components for 1 h. Three lengths of **I1** (5T, 10T and 15T) and **I2** (5T, 15T, 20T) were used in different combinations to obtain spacer lengths ranging from 10 to 35 nt. (T refers to thymine, which was the nucleotide used in the spacer). Credits to my FYP student, Mr Alex Sim Chen Siong, for performing this experiment

7.2.5 Detection of α-Thrombin Using Aptamers

Aptamer was first plugged directly into the existing circuit sequences as the recognition moiety for detecting α-thrombin. Thrombin is an important protease which regulates physiological coagulation [17]. Numerous aptamer based assays have been developed for thrombin detection and an excellent review is presented by Le et al. [18]. Due to its well-defined structure, we used α-thrombin to demonstrate the role of spacer (domain *s*) in modulating the separation distance of detection.

Three spacer lengths were incorporated into each initiator to obtain a combination of six separation distances. Some HCR products started to form from the separation distance of 20 nt onwards, and reached maximum signal intensity for 25–35 nt separation distances (Fig. 7.8). A total separation distance of 30 nt (15 nt on **I1** and 15 nt on **I2**) was used for subsequent α-thrombin works. Long HCR products formed unexpectedly even though equimolar amount of α-thrombin was used. Since the chain length of HCR product was inversely proportional to the effective target concentration, it suggests that the initiators bound to α-thrombin with low efficiency.

This could be due to the formation of undesired secondary structures between the aptamers and initiator domain sequences when the former was "plugged" directly into the existing circuit. (Figure 7.9). Instead of retrofitting the circuit, a new set of sequences was designed using the framework presented in Sect. 7.2.1, which took into consideration sequence constraints imposed by the aptamers. The revised sequence design preserved both the secondary structure of the aptamers and the single-stranded form of other domains for more efficient hybridization (Fig. 7.10).

The performance of the revised sequence design for α-thrombin detection was evaluated (Fig. 7.11a). The LOD of the new design was 0.5 nM (Fig. 7.11b) which was better than the 18.0 nM achieved using the old design (data not shown) or

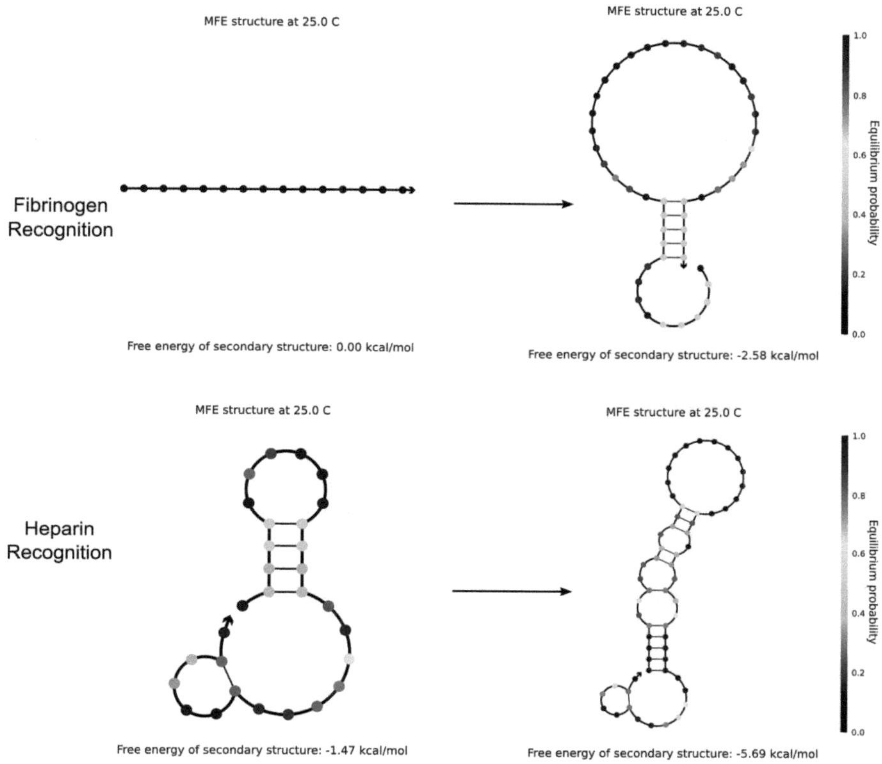

Fig. 7.9 Direct "plug-and-play" of aptamer sequences (left) into the original initiators design resulted in significant changes to the secondary structure (right)

the 50 nM reported for the first iteration of our circuit design (Sect. 4.2.4). The fibrinogen-recognizing aptamer has a relatively high dissociation constant of 100 nM [19], which means that only 0.2 nM was in bound form when 20 nM of **I2-apt15** reacted with 1 nM of α-thrombin. This was already at the LOD of our HCR FRET readout which currently stands at one-hundredth of the input hairpin concentration. The key bottleneck was likely the HCR readout module which failed to pick up signal from trace amount of thrombin-bound initiators.

7.2.6 Semi-quantitative Evaluation of Binding Affinity

The operation of the split proximity circuit is conditional upon the recognition of two binding events which are in close proximity. As such, it can potentially be used to evaluate the binding affinity of the recognition moiety to the target(s) or the binding affinity between the interacting pairs.

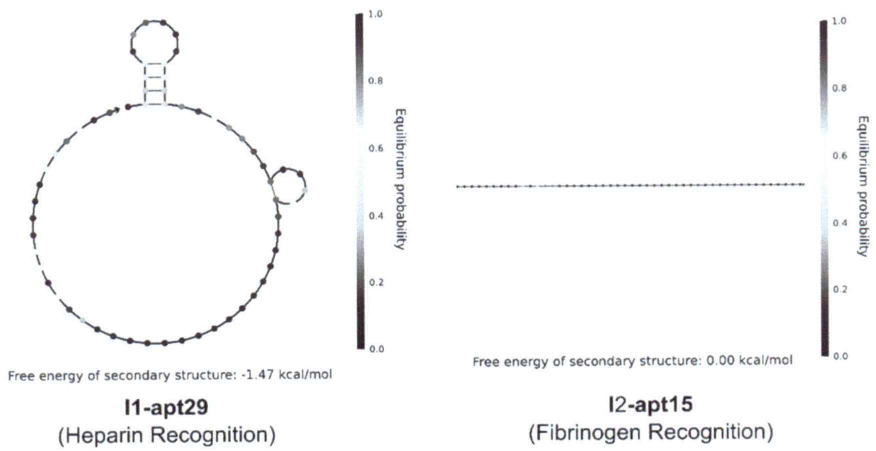

Fig. 7.10 Evolved design of initiators (**I1-apt29** and **I2-apt15**) for detecting thrombin. The original secondary structure of the aptamer was preserved while the circuit domains were available in single stranded form for more efficient hybridization

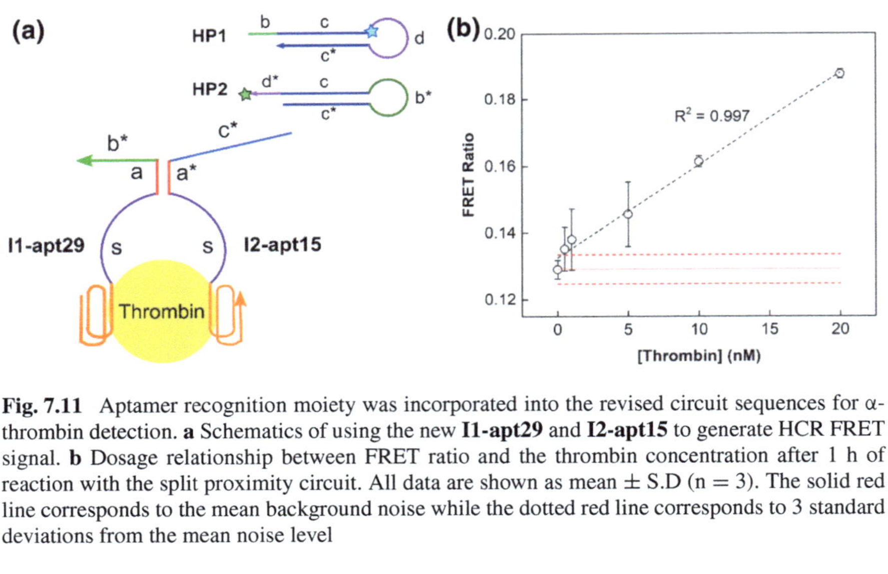

Fig. 7.11 Aptamer recognition moiety was incorporated into the revised circuit sequences for α-thrombin detection. **a** Schematics of using the new **I1-apt29** and **I2-apt15** to generate HCR FRET signal. **b** Dosage relationship between FRET ratio and the thrombin concentration after 1 h of reaction with the split proximity circuit. All data are shown as mean ± S.D (n = 3). The solid red line corresponds to the mean background noise while the dotted red line corresponds to 3 standard deviations from the mean noise level

In the first experiment, the binding affinity of **ST** to the combined recognition domains (x^* y^*) was investigated (Fig. 7.12a). Equimolar amount of **ST** was equilibrated with the split proximity circuit for an hour, followed by the titration of a perfectly complementary anti-**ST** (**aST**) strand to the complex. The displacement reaction was equilibrated overnight, and **F-Q** was added thereafter to characterize the concentration of remaining **ST-I1-I2** complex. A simple **F-Q** readout was pre-

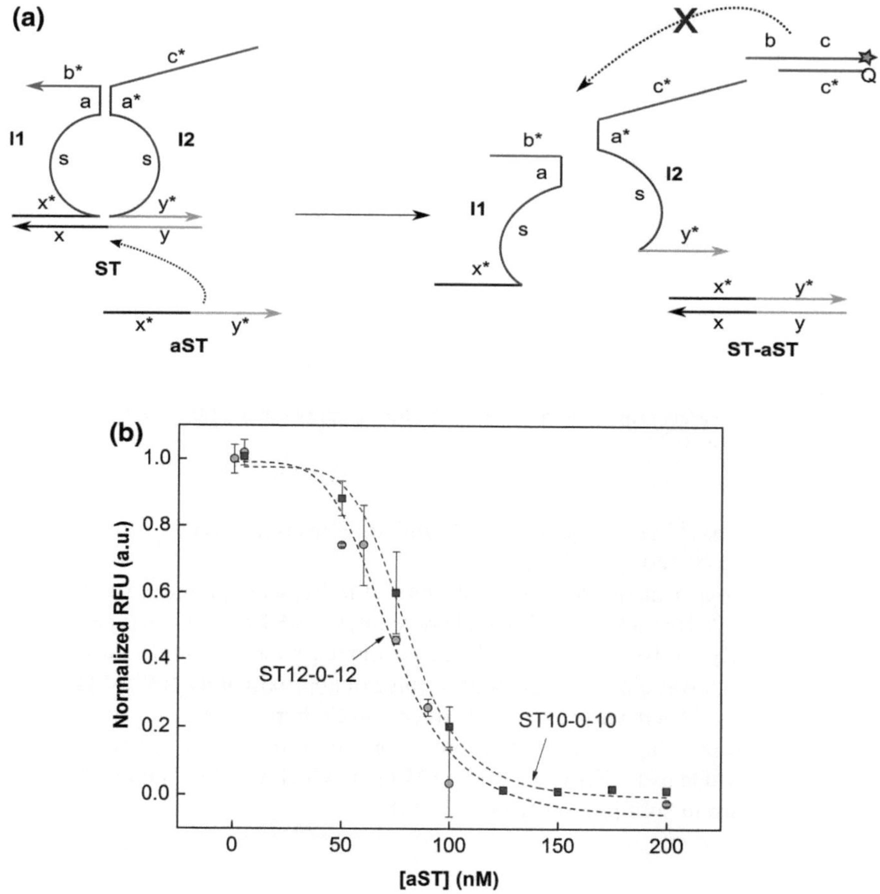

Fig. 7.12 The split proximity circuit was used to estimate the dissociation constant (K_D) of **ST** to the recognition domains. **a** Schematic outline of the titration experiment used to determine the dissociation constant of the binding between **ST** and its recognition moiety. **b** The higher the concentration of anti-ST (**aST**) added, the greater the disruption to the hybridization at domain *x* and *y* and hence the lower signal generated by the split proximity circuit. Two lengths of **ST** were used—12 nt (**ST12-0-12**) and 10 nt (**ST10-0-10**) for each domain. All data are shown as mean ± S.D (n = 3)

ferred over HCR for this application to avoid the complications arising from different degrees of HCR amplification on individual complexes.

The titration curve was fitted to a four-parameter logistic model commonly used to determine binding affinity [20]. The fitted dissociation constant (K_D) increased as the length of **ST** decreased, i.e. $K_{D,ST12012} = 72.9$ nM and $K_{D,ST10010} = 80.0$ nM, due to weaker DNA hybridization (Fig. 7.12b). The K_D obtained were of the same order

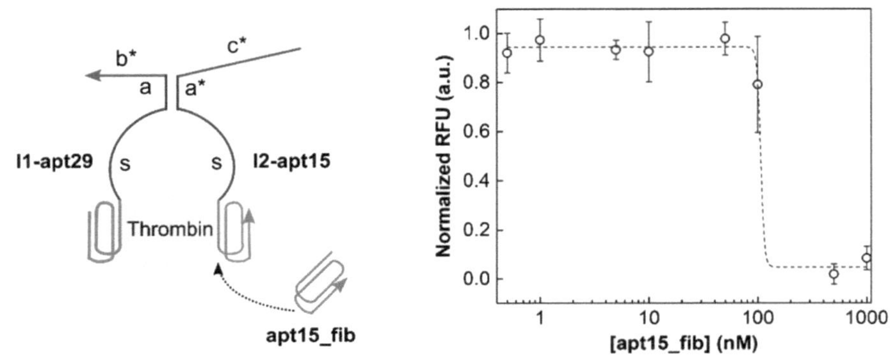

Fig. 7.13 Titration of increasing aptamer concentrations to the thrombin-circuit complex to deduce the dissociation constant (K_D) of fibrinogen recognition aptamer (**apt15_fib**). All data are shown as mean ± S.D (n = 3)

of magnitude (10^{-8} M) as sequences with similar Gibbs free energy of hybridization placed within confined space [21].

In the next experiment, the binding affinity of individual aptamer to α-thrombin was investigated. The same experimental procedures as **aST** titration was performed. We obtained K_D of 107 nM for the 15 nt fibrinogen exosite aptamer (**apt15_fib**) (Fig. 7.13) which was close to the literature value of approximately 100 nM [19]. The set-up involving 29 nt heparin exosite aptamer (**apt29_hep**) did not yield meaningful binding curve, possibly because the protein concentration used was not of similar order of magnitude as its K_D value of 0.5 nM [19], which was beyond the detection limit of our current split proximity circuit [22].

7.2.7 Detection of Endogenous HER2:HER2 Homodimer and HER2:HER3 Heterodimer

We evaluated the ability of our split proximity circuit to detect protein complexes under complex medium condition. The system under study was the human epidermal growth factor receptor (ErbB) family, which is implicated in cancer progression, particularly the HER2 biomarker which over expresses in 15–30% of breast cancers and is strongly associated with poor disease prognosis. There are no known ligands for HER2 which instead dimerizes with the other three members (HER1, HER3 and HER4) of the ErbB family or itself to trigger cell signalling [23].

Three cell lines were used for the detection of endogenous HER2 and HER3 dimerization: (1) AU565 breast cancer cell line which over expresses HER2 and HER3, (2) MDA-MB-231 triple negative breast cancer cell line with no HER2 and HER3 expression, and (3) MCF10A normal breast epithelial cell line with basal-to-no expression of HER2 and HER3. The membrane and cytosol proteins were

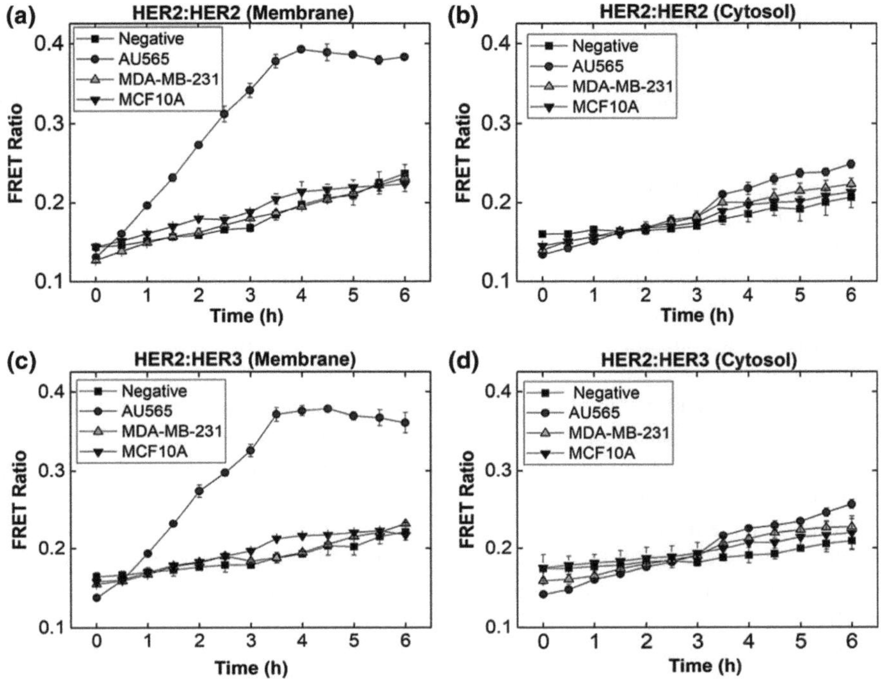

Fig. 7.14 Detection of HER2:HER2 homodimer in **a** membrane and **b** cytosol fraction, and HER2:HER3 heterodimer in **c** membrane and **d** cytosol fraction using the split proximity circuit. 100 μg/mL of the respective fractions extracted from AU565, MDA-MB-231 and MCF10A cell lines were used. Data are shown as mean ±S.D. (n = 3)

extracted from the cell lines in separate fractions as per manufacturer's protocol. The respective expression levels were confirmed using western blot (Appendix G). The cell lysate fractions were quantified and adjusted to 100 μg/mL across all fractions to ensure fairness of comparison. 100 μg/mL was ranged within 50–500 μg/mL and involved at least five times dilution recommended for commercially available cell lysate assays. No further purification of the cell lysate was carried out.

The HCR FRET developed appreciably higher signal for both HER2:HER2 homodimers and HER2:HER3 heterodimers in the membrane faction of the positive AU565 cell line than the negative MDA-MB-231 and normal MCF10A cells which were indistinguishable from the negative control where no target was added (Fig. 7.14). This shows that the circuit retained its specificity under a complex reaction environment consisting of the lysis buffer and a cocktail of extracted proteins. However, the kinetics was slower than under a controlled reaction environment and the equilibrium FRET ratio was achieved only after 3 h.

7.3 Conclusion

We successfully integrated the individual modules into a split proximity circuit driven by the reassembly of the central signal transduction module upon recognizing two binding events within close proximity. This chapter focused on developing the circuit for practical applications across different types of targets. The use of the circuit as a "plug-and-play" system was demonstrated—one may plug in the appropriate readout module to suit specific applications, or attach different types of recognition moiety corresponding to the target(s) of interest. Some applications covered in this chapter include end-point biomolecule detection, estimation of binding affinity between biomolecules and detection of protein dimers in a complex cell lysate environment.

Though the breadth of applications achievable using the split proximity circuit was shown, its performance fell short of satisfying all evaluation criteria of bio-assays, especially in terms of speed and LOD. The operational parameters still need to be fine-tuned for future applications. Well characterised targets were chosen as proof-of-concept systems in this chapter to confirm the functionality of our split proximity circuit with greater confidence. Moving on, we shall explore other more clinically relevant targets in the future given the positive results obtained in this preliminary study.

The strengths and limitations of the circuit were discussed throughout the application works. One recurring observation was that the HCR FRET module appeared to be a double-edged sword in the current design due to the chain termination when the trigger concentration reached one-hundredth of the input hairpin concentration. This mechanism either helped to suppress low levels of undesired reactions and hence enhance the circuit selectivity, or imposed a predictable poorer limit to the concentration of target(s) detected.

The split proximity circuit excelled in terms of good selectivity, robust operation and the simplicity of its one-pot execution. It can potentially value add to the plethora of existing detection platforms for applications where target(s) need to be fished out from a complex environment or where time-consuming procedures are still the norm. The detection of protein interactions at the cellular level, be it in lysate or whole-cell form, seems to fit the bill. We will look at a specific case study of detecting and visualizing HER2:HER2 and HER2:HER3 dimers on fixed cells in the next chapter.

References

1. Chen, X.: Expanding the rule set of DNA circuitry with associative toehold activation. J. Am. Chem. Soc. **134**, 263 (2012)
2. Nick Pace, C., Treviño, S., Prabhakaran, E., Martin Scholtz, J.: Protein structure, stability and solubility in water and other solvents. Philos. T. R. Soc. Lon. B **359**, 1225 (2004)
3. Formaneck, M.S., Ma, L., Cui, Q.: Effects of temperature and salt concentration on the structural stability of human lymphotactin: insights from molecular simulations. J. Am. Chem. Soc. **128**, 9506 (2006)

4. Owczarzy, R., Moreira, B.G., You, Y., Behlke, M.A., Walder, J.A.: Predicting stability of DNA duplexes in solutions containing magnesium and monovalent cations. Biochemistry **47**, 5336 (2008)
5. Cheglakov, Z., Cronin, T.M., He, C., Weizmann, Y.: Live cell MicroRNA imaging using cascade hybridization reaction. J. Am. Chem. Soc. **137**, 6116 (2015)
6. Dai, S., Xue, Q., Zhu, J., Ding, Y., Jiang, W., Wang, L.: An ultrasensitive fluorescence assay for protein detection by hybridization chain reaction-based DNA nanotags. Biosens. Bioelectron. **51**, 421 (2014)
7. Gao, Y., Wolf, L.K., Georgiadis, R.M.: Secondary structure effects on dna hybridization kinetics: a solution versus surface comparison. Nucleic Acids Res. **34**, 3370 (2006)
8. Leontis, N.B., Kwok, W., Newman, J.S.: Stability and structure of three-way DNA junctions containing unpaired nucleotides. Nucleic Acids Res. **19**, 759 (1991)
9. Fish, D.J., Horne, M.T., Searles, R.P., Brewood, G.P., Benight, A.S.: Multiplex SNP discrimination. Biophys. J. **92**, L89 (2007)
10. Ambros, V.: The functions of animal MicroRNAs. Nature **431**, 350 (2004)
11. de Planell-Saguer, M., Rodicio, M.C.: Analytical aspects of MicroRNA in diagnostics: a review. Anal. Chim. Acta **699**, 134 (2011)
12. Roush, S., Slack, F.J.: The let-7 family of MicroRNAs. Trends Cell Biol. **18**, 505 (2008)
13. Zhu, X., Shen, Y., Cao, J., Yin, L., Ban, F., Shu, Y., Li, G.: Detection of microrna snps with ultrahigh specificity by using reduced graphene Oxide-assisted rolling circle amplification. Chem. Commun. **51**, 10002 (2015)
14. Neely, L.A., Patel, S., Garver, J., Gallo, M., Hackett, M., McLaughlin, S., Nadel, M., Harris, J., Gullans, S., Rooke, J.: A single-molecule method for the quantitation of MicroRNA gene expression. Nat. Methods **3**, 41 (2006)
15. Tu, Y., Wu, P., Zhang, H., Cai, C.: Fluorescence quenching of gold nanoparticles integrating with a conformation-switched hairpin oligonucleotide probe for MicroRNA detection. Chem. Commun. **48**, 10718 (2012)
16. Cissell, K.A., Campbell, S., Deo, S.K.: Rapid, single-step nucleic acid detection. Anal. Bioanal. Chem. **391**, 2577 (2008)
17. Crawley, J.T.B., Zanardelli, S., Chion, C.K.N.K., Lane, D.A.: The central role of thrombin in hemostasis. J. Thromb. Haemost. **5**, 95 (2007)
18. Deng, B., Lin, Y., Wang, C., Li, F., Wang, Z., Zhang, H., Li, X.-F., Le, X.C.: Aptamer binding assays for proteins: the thrombin example—a review. Anal. Chim. Acta **837**, 1 (2014)
19. Rinker, S., Ke, Y., Liu, Y., Chhabra, R., Yan, H.: Self-assembled DNA nanostructures for distance-dependent multivalent ligand-protein binding. Nat. Nanotechonol. **3**, 418 (2008)
20. Prinz, H.: Hill coefficients, dose-response curves and allosteric mechanisms. J. Chem. Biol. **3**, 37 (2010)
21. Stevens, P.W., Henry, M.R., Kelso, D.M.: DNA hybridization on microparticles: determining capture-probe density and equilibrium dissociation constants. Nucleic Acids Res. **27**, 1719 (1999)
22. Hall, D.A., Langmead, C.J.: Matching models to data: a receptor pharmacologist's guide. Brit. J. Pharmacol. **161**, 1276 (2010)
23. Rubin, I., Yarden, Y.: The basic biology of HER2. Ann. Oncol. **12**, S3 (2001)

Chapter 8
DNA Split Proximity Circuit for Visualizing Cell Surface Receptor Clustering—A Case Study Using Human Epidermal Growth Factor Receptor Family

8.1 Introduction

Cell surface receptors play an important role in cellular signal transduction process, such as from the extracellular matrix to the cell [1] and to facilitate cell-to-cell communication [2]. A classic example is the T cell signalling pathway involved in immune response [3]. They also serve as important biomarkers for disease diagnosis [4, 5] and as therapeutics targets [6]. The spatial distribution of the receptors on the cell membrane regulates the clustering pattern and hence signalling process [7]. In particular, receptor dimerization and oligomerization often follow from ligand binding and are understood as the general mechanism for activating receptors with single transmembrane domain [8].

These receptor clustering events are typically detected in situ using advanced microscopy techniques capable of achieving nanometer-scale spatial mapping and temporal resolution in the range of seconds to milliseconds [9]. Examples of these imaging techniques are stimulated emission depletion (STED) microscopy [10] and stochastic optical reconstruction microscopy (STORM) [11]. These are useful techniques for understanding the fundamental dynamics of receptor clustering and interactions. However, they involve the use of sophisticated instruments and complex operations, and are not suitable tools for day-to-day operations.

In this chapter, we will apply our as-developed split proximity circuit for the one-pot detection and visualization of cell surface receptor clustering. The system under study is the human epidermal growth factor receptor (HER) family made up of four receptor members—HER1, HER2, HER3 and HER4, and controls cell growth and proliferation [12]. The receptors can exist as monomers and dimers, of which HER2 heterodimers trigger much stronger intracellular signalling than other dimer combinations [13]. As such, high expression level of HER2 is correlated with aggressive cell proliferation and tumour development in breast cancer, and is recognized as an important biomarker for the prognosis of breast cancer [14]. However, a significant proportion of patients show drug resistance against antibody-based therapy for HER2

A. Yan Shan, *Engineering a Robust DNA Circuit for the Direct Detection of Biomolecular Interactions*, Springer Theses,
https://doi.org/10.1007/978-981-13-2188-7_8

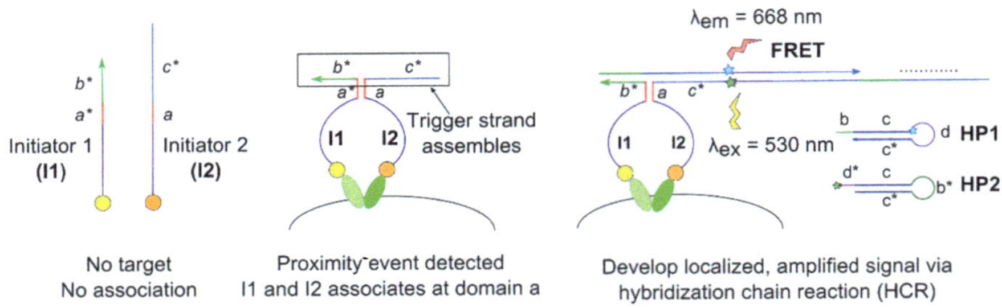

Scheme 8.1 Schematics of the split proximity circuit for the localized detection and visualization of cell surface receptor clustering

target [15]. It is believed that alternative signalling pathways activated, especially via HER2:HER3 heterodimerization, contribute to the drug resistance mechanism [16]. In this work, we will focus specifically on the detection and visualization of HER2:HER2 homodimers and HER2:HER3 heterodimers.

The basic circuit operation on cell surface is similar to that in solution-phase (Scheme 8.1). The initiator strands (**I1** and **I2**), conjugated to specific recognition antibodies, exist as separate entities in absence of the target receptors, or receptors which are not in close proximity. Once the circuit detects a pair of target receptors in close proximity, which we assume to be indicative of an interaction event, **I1** and **I2** associates to assemble the complete trigger strand. This triggers the hybridization chain reaction (HCR) readout module to generate localized, amplified fluorescent signal in the form of long DNA chains tagged with a Cy3-Cy5 FRET pair.

The key difference from the previous studies at the biomolecular level in Chap. 7 is that the circuit now operates on the surface of the cells which are anchored to a substrate, i.e. heterogeneous reaction [17]. This implies that parameters such as the spatial distribution and effective concentration of receptor targets, and the actual reaction volume, differ from the homogeneous solution-phase reaction. As such, we expect the circuit kinetics to be different from the pure biomolecular systems discussed previously.

The kinetics of our heterogeneous circuit reaction, i.e. efficiency of individual modules and the identification of rate-limiting step, were first characterized using standard cell enzyme-linked immunosorbent assay (ELISA). Once the circuit parameters were established, the detection of actual HER2:HER2 and HER2:HER3 clustering was attempted with the use of the split proximity circuit. We demonstrated that the one-pot format was feasible for cell analysis without incurring appreciable background noise. Finally, we showed that the formation of localized HCR FRET previously conceptualized could be used for the direct visualization of cell receptor clusters.

8.2 Results and Discussion

8.2.1 Characterizing Circuit Operation in Heterogeneous Reaction

The operation of the split proximity circuit was first validated using a standard cell ELISA format. This involved (i) the binding of antibody-conjugated initiator probes (**I1** and **I2**) to the fixed cells, (ii) thorough washing of unbound probes, (iii) addition of HCR hairpin probes (**HP1** and **HP2**) for signal development and (iv) final washing steps to remove unbound hairpins (Scheme 8.2). The step-by-step approach in the standard method allowed us to quantify the heterogenous reaction of the circuit in terms of (1) efficiency of the three-way junction in transducing surface-bound recognition events into readout signals, i.e. signal transduction efficiency, and (2) efficiency of utilizing HCR hairpins for heterogeneous reactions, i.e. HCR efficiency.

We first confirmed that the circuit was operating as designed on cell systems by introducing different initiator concentrations to detect HER2:HER2 homodimers on AU565 cells which highly expresses HER2 and HER3 (Appendix G). The FRET signal generated via HCR was monitored real-time after the addition of the hairpin monomers. Clear dosage dependence of FRET signal on the initiator concentrations was observed (Fig. 8.1). There was appreciable background noise, which gradually grew over time, in absence of initiators. This was in excess of the circuit leakage observed when all circuit components were present without the target cell. The observations suggest that the background noise was likely due to non-specific adsorption of the DNA-based hairpin monomers on the cell surface while the circuit itself remained specific in its operation.

The gradual drop in FRET ratio in absence of cells was due to inherent instability of the internally modified Cy5 molecule which fluorescence intensity tended to decrease over time. We also note the large standard deviations obtained from triplicate experiments which may be attributed to the heterogeneous nature of cell-based assay with variations from well to well and from receptor expression of individual cells within each well.

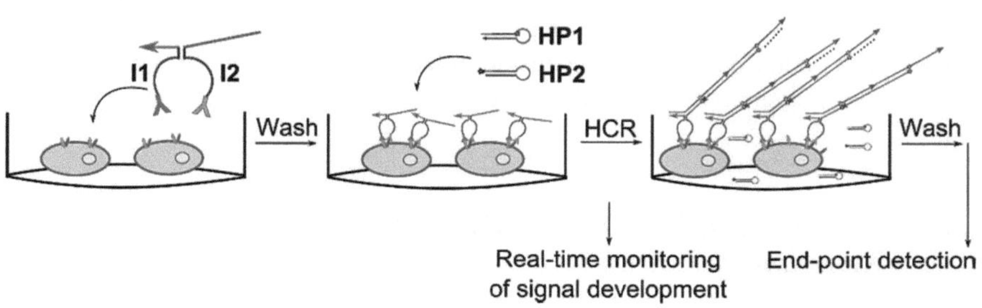

Scheme 8.2 Schematics of the standard cell ELISA procedures except that hybridization chain reaction (HCR) was used for signal development

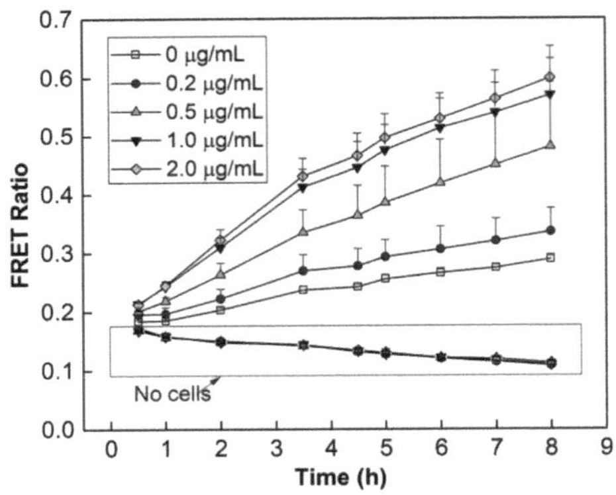

Fig. 8.1 The kinetics of the circuit operation was monitored real-time using the HCR FRET readout signal. Five concentrations of anti-HER2-**I1** and anti-HER2-**I2** (0–2.0 μg/mL), together with 40 nM **HP1** and 20 nM **HP2**, were added to the cells. The development of FRET signal was dependent on both initiator concentrations and reaction time. Circuit leakage was minimal in absence of cell target (marked out in red rectangle and labelled with "no cells"). Data are presented as mean ± standard deviation of triplicate experiments

The circuit kinetics was noticeably slower than in solution phase. We had previously characterized the reaction rate constant of the circuit as 2.14×10^4 M^{-1} s^{-1} ($\tau_{reaction}$ ≈27 min) in Sect. 7.2.3. The diffusion coefficient (D) of a 30 nt oligonucleotide is approximately 9.0×10^{-7} cm^2/s [18], while that of lysozyme (whose molecular weight of 14 kDa is equivalent to the HER2 affibody) is approximately 1.1×10^{-6} cm^2/s [19]. We used the slower of the two, i.e. 9.0×10^{-7} cm^2/s, for subsequent calculations. A back of envelope estimation of the characteristic diffusion time for the circuit components to diffuse from the mid-point of the total reaction volume to the cells at the bottom of the well was approximately 1.5 h, assuming three-dimensional random walk (Fig. 8.2a). Since the characteristic time of diffusion was approximately three times slower that of reaction, we deduced that the circuit operation in heterogeneous reaction was diffusion limited. This deduction was further supported by the linear relationship between the FRET signal and square root of time which was indicative of a diffusion limited reaction (Fig. 8.2b) [20]. We should point out that the above analysis provided only a simplified picture of the actual circuit operation and did not consider intermediate reaction steps, e.g. kinetics of antibody-antigen binding, and the use of complex buffer environment.

8.2.2 *Efficiency of Split Proximity Circuit in Heterogeneous Reaction*

The circuit operation as a cell-based assay was characterized via measuring Cy5 fluorescence intensity. The signal transduction efficiency was approximated by com-

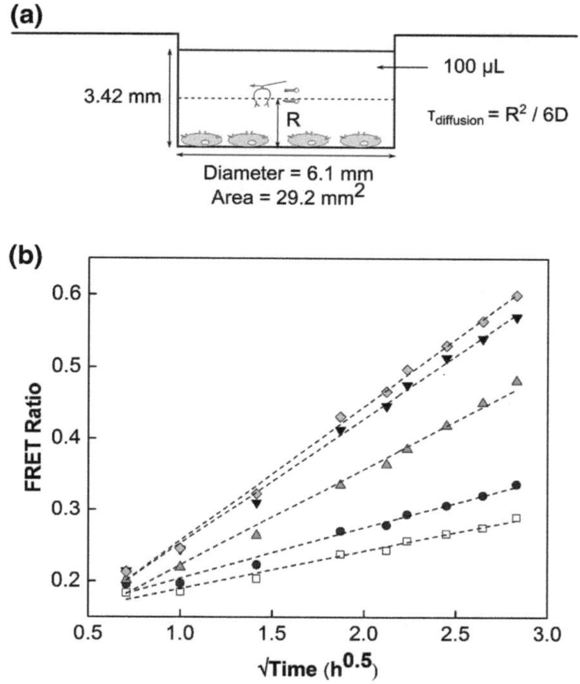

Fig. 8.2 Estimation of mass transfer parameters in the heterogeneous reaction involving the split proximity circuit on cell surfaces. **a** Illustration of the key parameters involved in the 96-well ELISA used to calculate the characteristic diffusion time for the circuit components to diffuse from the mid-point (R) of reaction volume to cells at the bottom of the well. **b** The FRET signal evolution was linearly proportionate to the square root of time, which indicated that the circuit operation was diffusion limited. The best fit lines (dotted) returned R^2 values between 0.98 and 0.99

paring to the case where the direct trigger (**DT**) for HCR was used in place of the split initiators strand (**I1** and **I2**) (Fig. 8.3a). In this case, **DT** was conjugated to anti-HER2, while the split proximity circuit was used to detect HER2:HER2 homodimer in AU565 cells.

As the concentration of the initiators increased from 0.2 to 1.0 μg/mL, the signal transduction efficiency increased (Fig. 8.3b). This suggests that the initiators were in limiting concentration and so a pair of anti-HER2-**I1** and anti-HER2-**I2** might not be simultaneously present in close vicinity to detect the HER2:HER2 homodimer. Beyond 1.0 μg/mL, the signal transduction efficiency appeared to stabilize at approximately 70–75%.

It should be noted that **DT** would trigger HCR in response to the recognition of a single HER2 molecule, be it in monomeric or dimeric (HER2:HER1/2/3/4) form, and overestimate the maximum FRET signal achievable (Fig. 8.4a). On the contrary, it was possible for the circuit to be falsely triggered in presence of neighbouring pairs of HER2 but not necessarily existing as true dimers (Fig. 8.4b). This limitation

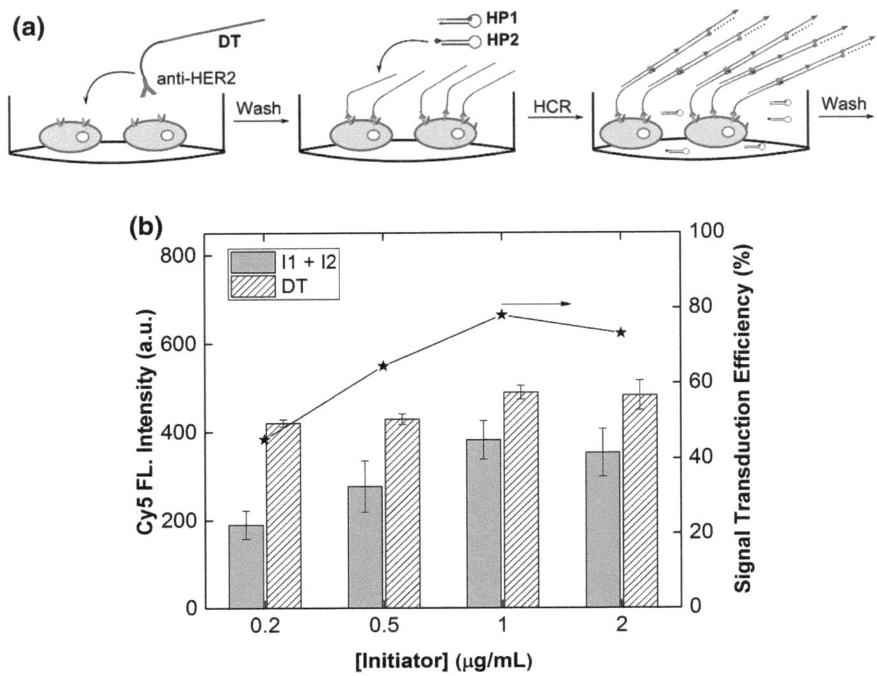

Fig. 8.3 Characterising signal transduction efficiency of the split proximity circuit. **a** Schematics of the standard cell ELISA procedures performed using the direct trigger (**DT**) strand for HCR instead of the split initiators strands. **b** By comparing the Cy5 fluorescence (FL.) intensity when **I1** and **I2** were used in the split proximity circuit versus **DT** after washing, the signal transduction efficiency was approximated. Data are presented as mean ± standard deviation of triplicate experiments

is more prominent for the scenario where the target receptors are highly expressed, e.g. AU565 in this case. It is also a common problem for existing molecular toolbox, e.g. fluorescence co-localization [21]. We reiterate that the circuit was designed to recognize biomolecular events in close proximity with the inherent assumption that they are indicative of interaction events. One way to minimize the false detection of clustering events is to reduce the spacer length of the initiators to constrain the effective radius of detection.

We next evaluated the efficiency of HCR in heterogeneous reaction by comparing the intensity of Cy5 fluorescence before and after washing for the **DT** system (Fig. 8.5a). The act of washing removed unreacted hairpin monomers and those that were non-specifically bound either to the well or cell surface. We observed that the Cy5 fluorescence intensity remained relatively constant across the entire range of anti-HER2-**DT** concentrations used and the HCR efficiency appearing to be capped at approximately 75% (Fig. 8.5b). It was likely that the bulk of the unreacted hairpin monomers (approximately 25% of input hairpins) were non-specifically bound and subsequently washed away. The other hairpins in bulk solution effectively partici-

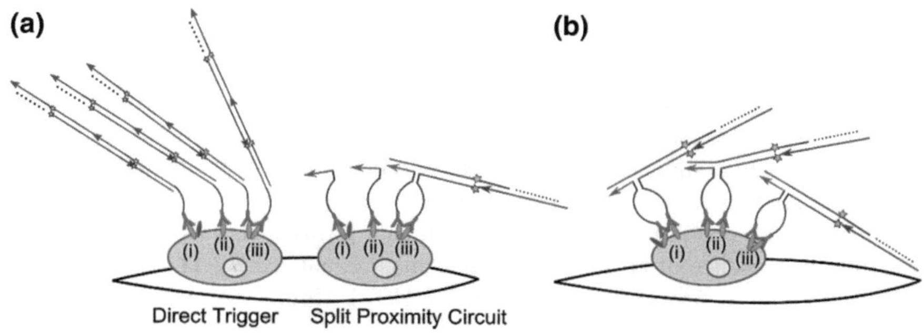

Fig. 8.4 Examples of false positive signals when applying the split proximity circuit on cell models. **a** The signal from the direct trigger was reflective of HER2 expression, be it in (i) heterodimer form, (ii) monomeric form or (iii) HER2:HER2 homodimer. On the other hand, the complete split proximity circuit was triggered only in presence of HER2:HER2 homodimers. **b** The split proximity circuit was designed to detect two targets in close physical proximity which might not necessarily be due to actual interactions. Both (i) cross talk between dimers and (ii) closely-packed monomers could falsely trigger the circuit besides the (iii) true HER2:HER2 homodimers

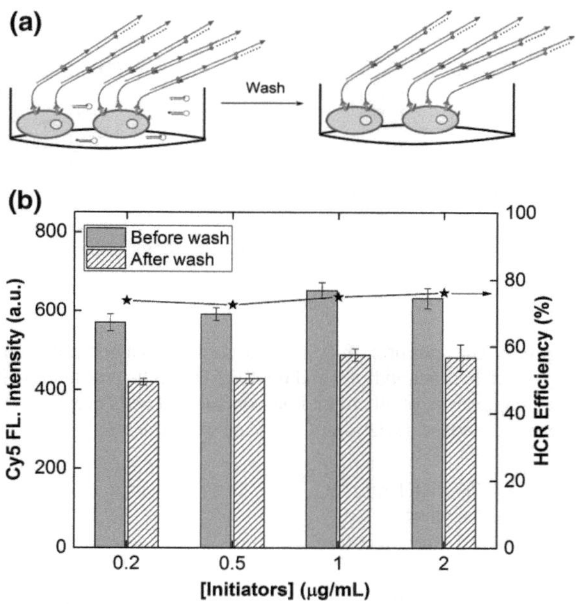

Fig. 8.5 Characterising HCR efficiency of the split proximity circuit. **a** The unbound HCR hairpins were washed in the final step of the standard cell ELISA. **b** The comparison of the Cy5 FL. intensity before and after cell washing gave an estimate on the efficiency of utilizing the HCR hairpins (HCR efficiency, %). Data are presented as mean ± standard deviation of triplicate experiments

pated in HCR and so, the reaction was likely to be limited by the hairpin concentration. It may be possible to boost the signal by increasing the amount of input hairpins, but this was not tested since it would lead to increased background noise.

8.2.3 One-Pot Detection of HER2:HER2/3 Dimers

After confirming that the circuit operated as per designed in heterogeneous reaction, we tested the performance of the split proximity circuit as a one-pot assay. In this case, the antibody-conjugated initiators and HCR hairpins were pre-mixed and added together to AU565 cells without intermediate washing steps. The circuit remained robust under the one-pot format where significant positive signals developed over time only in presence of all system components, i.e. anti-HER2-**I1**, anti-HER2-**I2**, **HP1**, **HP2** and AU565 cell, for both HER2:HER2 (Fig. 8.6a) and HER2:HER3 dimers (Fig. 8.6b).

While the circuit operation had yet to reach equilibrium after 4 h (Fig. 8.6), the FRET ratio for the test was already statistically more significant ($p < 0.005$) than all other controls at the first measurement time point of 30 min (Fig. 8.7). The negative control was taken as the entire split proximity circuit without the antibody recognition moiety, which accounted for the inherent circuit leakage. The lack of appreciable signal upon the omission of either anti-HER2-**I1** and anti-HER2-**I2** confirmed that the circuit was triggered only when two biomolecular recognition events occurred.

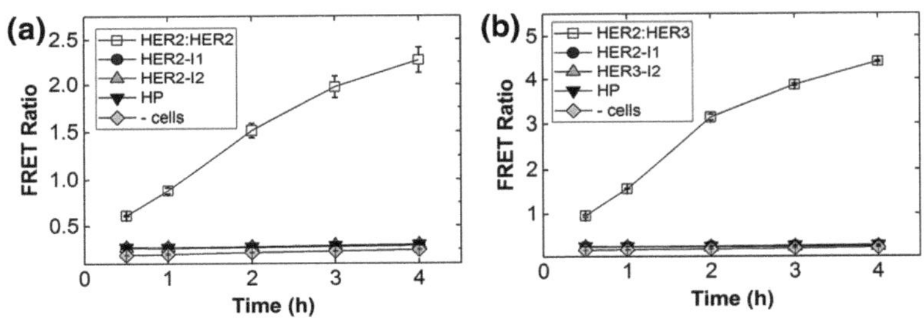

Fig. 8.6 The split proximity assay was added in a one-pot manner to seeded AU565 cells to detect **a** HER2:HER2 homodimer and **b** HER2:HER3 heterodimer. In both cases, significant signal developed in presence of all circuit components and cells over the control set-ups. Data are presented as mean ± standard deviation of triplicate experiments

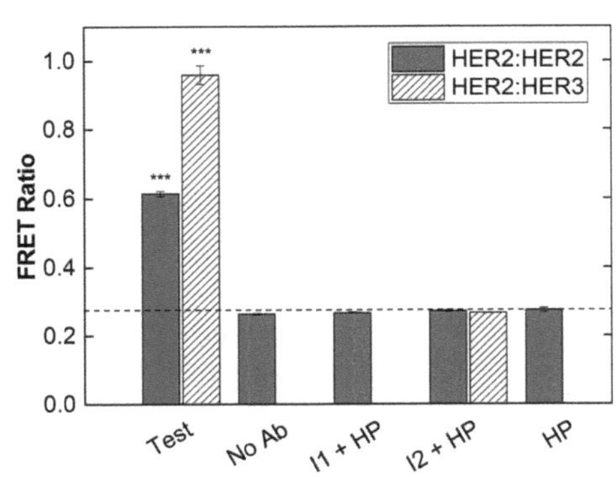

Fig. 8.7 The FRET ratio of the test and control set-ups was analyzed after 30 min of reaction. One-sided t-test was performed for test samples against the technical control where no antibody was conjugated to the initiator strands (no Ab). ***refers to $p < 0.005$ (n = 3)

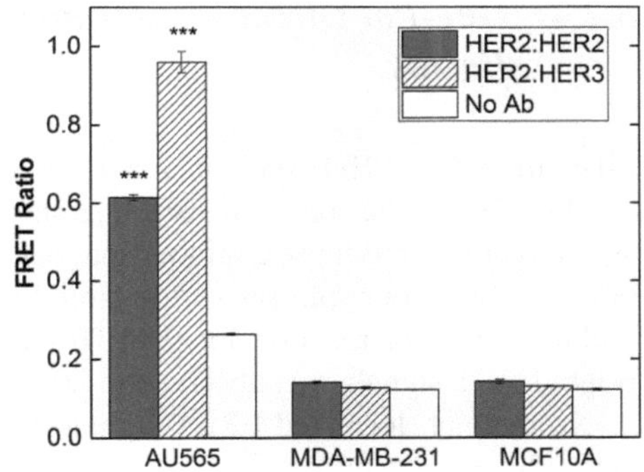

Fig. 8.8 The one-pot split proximity circuit was applied on all three cell lines: AU565, MDA-MB-231 and MCF10A. FRET was measured after 30 min of reaction. ***refers to $p < 0.005$ for one-tailed t-test ($n = 3$)

The detection time of the split proximity circuit in its current form was on the scale of hours, with the shortest feasible time being 30 min. As such, it is unable to probe for the dynamics of receptor clustering in real time if they occur at a time scale faster than the circuit operation. Instead, it can only be used to profile strong interactions with long binding time, e.g. 30–60 min of binding between T-cell and antigen-presenting cells [22], or interactions at a particular time point within fixed samples.

The one-pot split proximity circuit was next tested on the negative (MDA-MB-231) and normal (MCF10A) cell lines. The FRET ratio remained low and indistinguishable from the negative control (no antibody) for both MDA-MB-231 and MCF10A (Fig. 8.8), which was within expectation since both cell lines had negligibly low HER2 and HER3 expression. This proved that the split proximity circuit was specific in its operation and could be used to differentiate cells with different surface marker profiles.

One should note that the circuit recognition specificity is largely dependent on the specificity of the antibody used. Another factor would be the varying stickiness of different cell surface which could lead to varying degree of non-specific binding of the circuit components on the cells. In this case, AU565 could be a stickier cell than MDA-MB-231 and MCF10A cells which accounted for the higher FRET ratio for its negative control. However, such non-specific binding was less of an issue in our split proximity circuit than in typical ELISA assay since it required two different initiator strands to be non-specifically bound in close proximity to each other and with proper orientation to trigger HCR. Hence, intermediate washing steps were not critical to the success of our assay though one could choose to perform a single washing step at the end of reaction to reduce variations in the results.

8.2.4 One-Pot Direct Visualization of Receptor Clustering Events

Since the HCR FRET readout signal generated a long fluorescent, DNA chain that was localized at the point of detection, it should be possible to visualize the interaction sites directly. Time-lapse confocal imaging was performed on AU565 cells after one-pot addition of our split proximity circuit (Fig. 8.9). Visible signals started to form within 10 min of reaction for both HER2:HER2 and HER2:HER3 detection, and bright FRET signals was observed in 30 min.

No distinct, localized FRET signal was seen for the negative control (with no antibody conjugated), though a few stray fluorescent dots could be seen moving around the viewing frame over time. These could be HCR products mis-triggered in solution, or aggregation of hairpins non-specifically bound to the BSA protein in the reaction buffer. The time-lapse study further confirmed that 30 min was sufficient for generating sufficient signal over background noise as previously concluded from the ELISA results. All analysis from this point onwards was fixed at 30 min reaction time.

The specificity of the circuit operation was confirmed by omitting different circuit components in a series of controls (Fig. 8.10). Signal did not develop in the negative control where the circuit lacked the antibody recognition moiety. Neither was there signal when only one of the initiators was used which again confirmed that two recognition events were needed as inputs to the circuit. Lastly, no signal was observed when only the HCR hairpins were added which suggests that non-specifically bound hairpins were too diffuse on the cell membrane to be seen visibly under our confocal settings.

Having established the speed and specificity of circuit operation for direct visualization, we proceeded to test its ability to differentiate between various surface receptor profiles on different cell lines under confocal imaging. Again, strong FRET signal developed only for HER2:HER2 and HER2:HER3 dimers in AU565 cells (Fig. 8.11). The other two cell lines with no HER2 and HER3 expression did not develop appreciable FRET signal. Similarly, the negative controls returned no FRET signal.

8.3 Conclusion

Taken in its entirety, the split proximity circuit could perform its operation autonomously in heterogeneous reaction. The signal transduction efficiency was about 80% of that in solution-phase while HCR readout could proceed to completion. While the circuit operation remained effective, the heterogenous reaction was now diffusion limited and the effective assay time scale was in the range of hours. Nonetheless, the circuit successfully operated with good specificity and minimal circuit leakage to produce significant signal over background noise within 30 min and

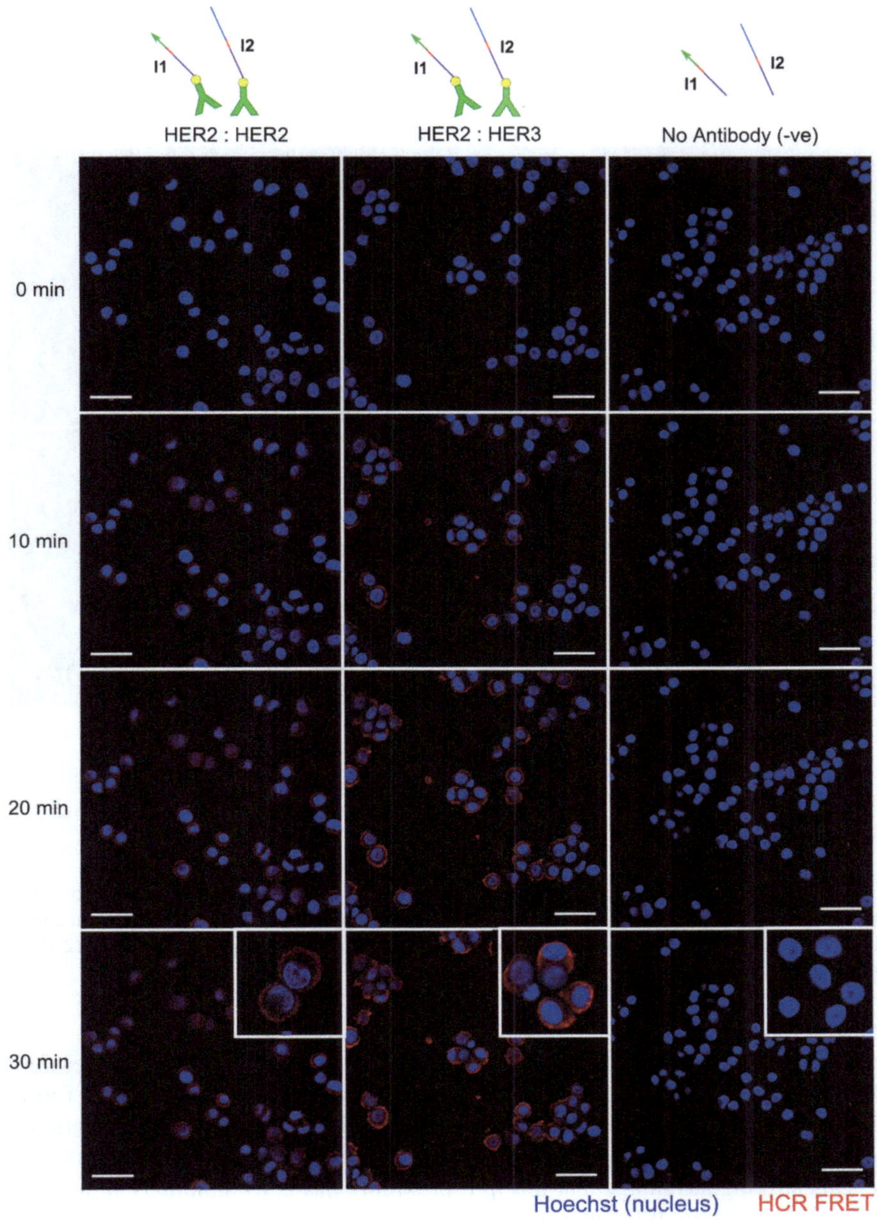

Fig. 8.9 Confocal images at various time points (0, 10, 20 and 30 min) after the one-pot split proximity circuit was added to AU565 cells to probe for HER2:HER2 homodimers (first column) and HER2:HER3 heterodimers (second column). Negative control where no antibody was added is shown in the third column. An enlarged image of the cells is shown in the insert of the confocal image at 30 min. Scale bar shown corresponds to 50 μm

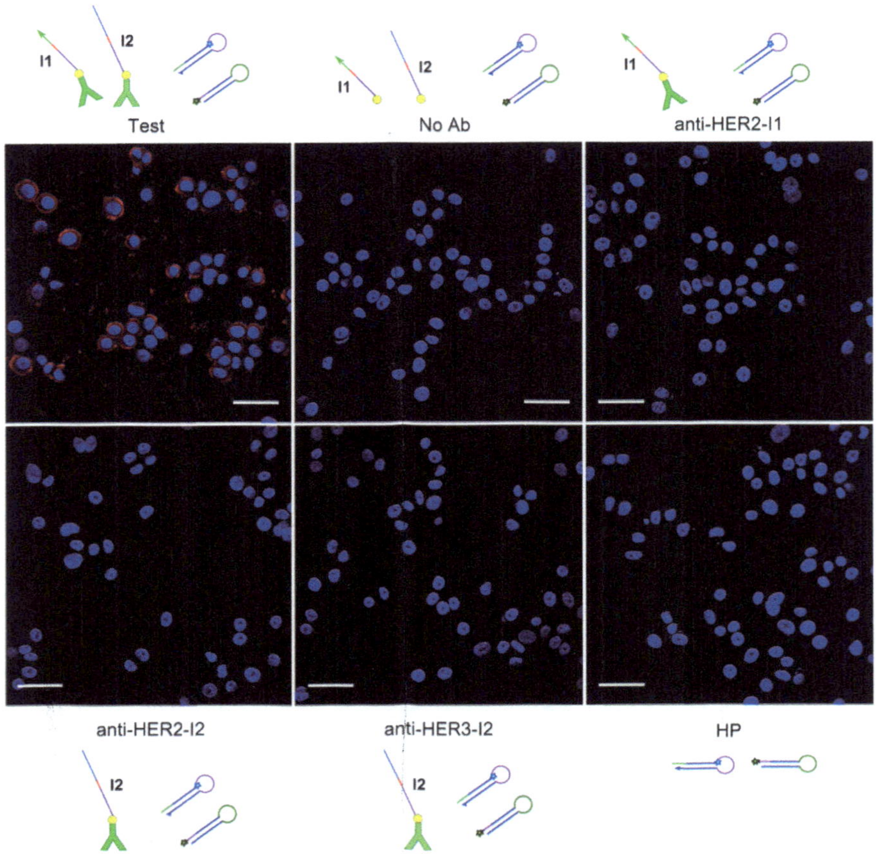

Fig. 8.10 The specificity of the components in the one-pot split proximity circuit was confirmed under confocal microscopy visualization. The circuit was added to AU565 cells in different combinations and visualized after 30 min of reaction. The positive test image corresponds to HER2:HER3 heterodimer. Scale bar shown corresponds to 50 μm

in absence of washing steps. Circuit computation on the cell surface successfully differentiated different receptor expression profiles on three cell types. The formation of localized, amplified HCR FRET signal facilitated the direct visualization of receptor clustering in situ.

One striking advantage that the split proximity has is its simplicity in terms of end-point operation. While the detailed working mechanism of our circuit may seem complicated, the final form of the assay can operate autonomously based on the inputs detected. Its one-pot format reduces processing time, laborious work and more importantly, cuts down the chance of human errors. Also, our assay is compatible with existing bio-analytical methods, *i.e.* antibody recognition, ELISA and confocal imaging.

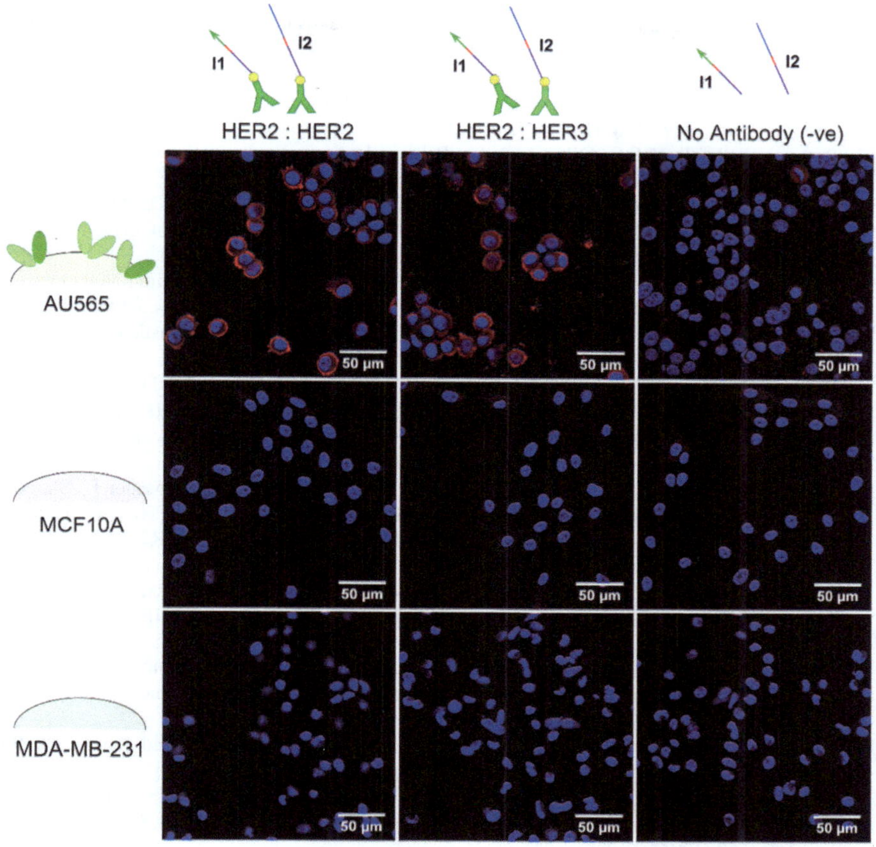

Fig. 8.11 Direct visualization of HER2:HER2 homodimer and HER2:HER3 heterodimer under confocal microscopy after 30 min of incubation in three cell lines: AU565 (high expression), MCF10A (normal, basal-to-no expression) and MDA-MB-231 (triple negative breast cancer cell, no expression)

References

1. Juliano, R.L., Haskill, S.: Signal transduction from the extracellular matrix. J. Cell Biol. **120**, 577 (1993)
2. Singer, S.J.: Intercellular communication and cell-cell adhesion. Science **255**, 1671 (1992)
3. Cochran, J.R., Aivazian, D., Cameron, T.O., Stern, L.J.: Receptor clustering and transmembrane signaling in T cells. Trends Biochem. Sci. **26**, 304 (2001)
4. Gedye, C.A., Hussain, A., Paterson, J., Smrke, A., Saini, H., Sirskyj, D., Pereira, K., Lobo, N., Stewart, J., Go, C., Ho, J., Medrano, M., Hyatt, E., Yuan, J., Lauriault, S., Kondratyev, M., van den Beucken, T., Jewett, M., Dirks, P., Guidos, C.J., Danska, J., Wang, J., Wouters, B., Neel, B., Rottapel, R., Ailles, L.E.: Cell surface profiling using high-throughput flow cytometry: a platform for biomarker discovery and analysis of cellular heterogeneity. PLoS ONE **9**, e105602 (2014)

5. Joensson, H.N., Samuels, M.L., Brouzes, E.R., Medkova, M., Uhlén, M., Link, D.R., Andersson-Svahn, H.: Detection and analysis of low-abundance cell-surface biomarkers using enzymatic amplification in microfluidic droplets. Angew. Chem. Int. Ed. **48**, 2518 (2009)
6. Christopoulos, A.: Allosteric binding sites on cell-surface receptors: novel targets for drug discovery. Nat. Rev. Drug Discov. **1**, 198 (2002)
7. Bethani, I., Skånland, S.S., Dikic, I., Acker-Palmer, A.: Spatial organization of transmembrane receptor signalling. EMBO J. **29**, 2677 (2010)
8. Heldin, C.-H.: Dimerization of cell surface receptors in signal transduction. Cell **80**, 213 (1995)
9. Valley, C.C., Lidke, K.A., Lidke, D.S.: The spatiotemporal organization of ErbB receptors: insights from microscopy. Cold Spring Harb. Perspect. Biol. **6** (2014)
10. Hein, B., Willig, K.I., Wurm, C.A., Westphal, V., Jakobs, S., Hell, S.W.: Stimulated emission depletion nanoscopy of living cells using SNAP-Tag fusion proteins. Biophys. J. **98**, 158 (2010)
11. Rust, M.J., Bates, M., Zhuang, X.: Sub-diffraction-limit imaging by stochastic optical reconstruction microscopy (STORM). Nat. Methods **3**, 793 (2006)
12. Rubin, I., Yarden, Y.: The basic biology of HER2. Ann. Oncol. **12**, S3 (2001)
13. Yarden, Y.: Biology of her2 and its importance in breast cancer. Oncology **61**(suppl 2), 1 (2001)
14. Weigel, M.T., Dowsett, M.: Current and emerging biomarkers in breast cancer: prognosis and prediction. Endocr.-Relat. Cancer **17**, R245 (2010)
15. Vogel, C.L., Cobleigh, M.A., Tripathy, D., Gutheil, J.C., Harris, L.N., Fehrenbacher, L., Slamon, D.J., Murphy, M., Novotny, W.F., Burchmore, M., Shak, S., Stewart, S.J., Press, M.: Efficacy and safety of trastuzumab as a single agent in first-line treatment of HER2-overexpressing metastatic breast cancer. J. Clin. Oncol. **20**, 719 (2002)
16. Sergina, N.V., Rausch, M., Wang, D., Blair, J., Hann, B., Shokat, K.M., Moasser, M.M.: Escape from HER-family tyrosine kinase inhibitor therapy by the kinase-inactive HER3. Nature **445**, 437 (2007)
17. Butler, J.E.: Immunochemistry of Solid-Phase Immunoassay. Taylor & Francis (1991)
18. de la Torre, G., Navarro, J., Navarro, S., Martinez, M.C.L.: Hydrodynamic properties of a double-helical model for DNA. Biophys. J. **66**, 1573 (1994)
19. Brune, D., Kim, S.: Predicting protein diffusion coefficients. Proc. Natl. Acad. Sci. U.S.A. **90**, 3835 (1993)
20. Nygren, H., Stenberg, M.: Kinetics of antibody-binding to surface-immobilized antigen: influence of mass transport on the enzyme-linked immunosorbent assay (Elisa). J. Colloid Interface Sci. **107**, 560 (1985)
21. Dunn, K.W., Kamocka, M.M., McDonald, J.H.A.: Practical guide to evaluating colocalization in biological microscopy. Am. J. Physiol.-Cell Ph. **300**, C723 (2011)
22. Germain, R.N.: T-cell signaling: the importance of receptor clustering. Curr. Biol. **7**, R640 (1997)

Chapter 9
Conclusion and Future Outlooks

9.1 Conclusion

This doctoral work culminated in a circuit design capable of detecting bi-molecular proximity events, which we named "split proximity circuit". The circuit assembles dynamically on the target recognition sites in a self-contained manner to execute localized strand displacement. The rationale behind the circuit design was driven by the fundamental concept of enhanced strand displacement kinetics due to the increased local DNA concentration from the localization effect. As such, the circuit operates only when three inputs were simultaneously detected—two target binding events which occur within close proximity, to achieve one-pot, turn-on detection.

A recurring theme in this thesis was to minimize circuit leakages and a major contribution of this thesis was the development of several key counter-leakage strategies. First, a four-point design guideline for obtaining robust hybridization chain reaction (HCR) hairpin was established and validated for 15 sets of sequences. With this, we flexibly re-designed hairpin sequences with shorter stem length and hence faster hairpin opening kinetics without incurring leakages. A distance-sensitive FRET-based readout was also incorporated into HCR. Next, we provided direct experimental evidence that synthesis defect at the 3′-end of oligonucleotide was the main culprit of truncated sequences. Finally, this inspired an "inter-domain bridging" strategy to counter the defect and boost strand displacement kinetics across a three-way junction.

We believe the strategies highlighted above are useful for designing DNA circuits in general and particularly as a starting point for researchers outside the DNA circuitry field. Within the scope of my doctoral studies, these design strategies were crucial in ensuring that the integrated modules in our final split proximity circuit continued to operate robustly under different environments tested, i.e. buffered condition, 10% FBS and cell lysate.

The greatest advantage of our split proximity circuit, when compared to existing detection technologies, is the operation simplicity. Once the design of the circuit operation is optimized, the final assay can be used in a one-pot format to oper-

A. Yan Shan, *Engineering a Robust DNA Circuit for the Direct Detection of Biomolecular Interactions*, Springer Theses,
https://doi.org/10.1007/978-981-13-2188-7_9

ate autonomously based on the inputs detected. Practically, this reduces processing time, laborious work and reduce chances of human error. However, the current circuit design is only able to sense at the biomolecular level or protein complexes present in close proximity and has yet to be validated as a functional assay capable of assessing true interaction events. Also, the time scale of circuit operation is in the order of minutes to hour and cannot be used to probe the dynamics of fast biomolecular interactions. The greatest potential of our as-developed circuit is for use as a diagnostics tool in clinical settings or as a screening platform to evaluate the performance of drug candidates and multivalent ligands in regulating protein-protein and cell surface interactions. [1, 2] We have immediate plans to develop the split proximity circuit as a diagnostics or imaging tool for the clinical evaluation of breast cancer biomarkers.

Another striking feature of our circuit design is its "plug-and-play" versatility where the corresponding recognition moiety can be used to detect specific target of interest. The proof-of-concept systems demonstrated in this thesis include end-point detection of single nucleotide mutations in DNA and RNA, estimation of binding affinity between α-thrombin and its recognition aptamers, and the visualization of clustering events between cell surface receptors. While each application has yet to be investigated in depth for specific biological or clinical question, the preliminary results showcased the potential of this platform as an enabling tool for various end applications, e.g. screening of drugs to disrupt biomolecular interactions, [3, 4] enhancing the understanding of biological interaction networks [5, 6] and as a monitoring tool for cancer prognosis [7–9].

9.2 Recommendations for Future Works

9.2.1 Modifications to Individual Split Proximity Circuit Modules

The individual modules of our split proximity circuit can be further expanded (Scheme 9.1). First, the system can be optimized for real-time analysis of live cells. For a start, the 6 nt association region design was functional under physiological conditions (Sect. 7.2.2). The downstream cellular responses of live cells can be used to characterize the functional effects of clustering events. Successful validation on live cell models should promote future development of the circuit as a functional assay for applications such as drug screening.

One potential challenge may be the stability of the DNA circuit under physiological condition due to the presence of nucleases. Different oligonucleotide modifications, e.g. using phosphorothioate bonds and inverted dT and ddT, can be included to reduce the effect of exonuclease degradation. Nucleic acid analogs, e.g. locked nucleic acid (LNA), peptide nucleic acid (PNA) and morpholino, possess improved stability against nuclease degradation and higher binding affinity than DNA molecules [10]. In particular, PNA and morpholino are non-ionic which may reduce

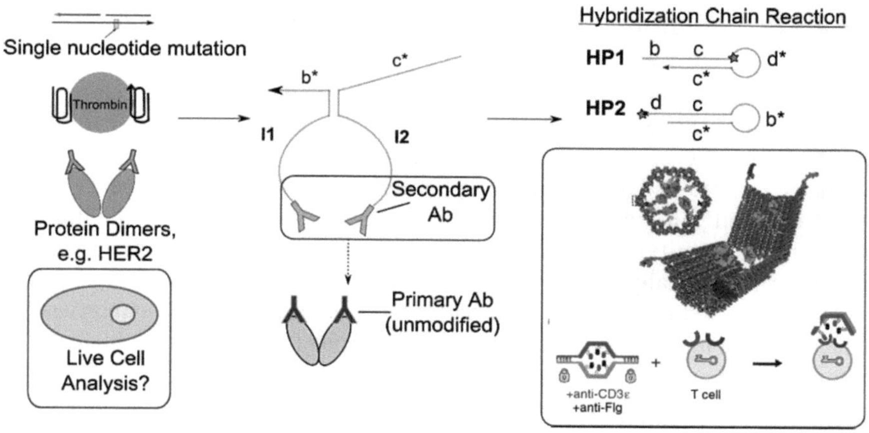

Scheme 9.1 Expanding on individual modules of the split proximity circuit: (From left to right) Detection of live cell, using secondary antibody (Ab) as a more generic recognition moiety, and incorporating a functional output. The DNA origami cage design was reproduced with permission from Ref. [14] [Reprinted with permission from AAAS]

the sensitivity of circuit performance to the ionic strength of the reaction buffer. However, their stronger binding affinity also implies potentially reduced sequence recognition specificity [11]. A thorough study is needed to evaluate the trade-offs.

Next, the generality of the split proximity circuit can be expanded by using secondary antibody as the recognition moiety (current design is based on primary antibody recognition). In this way, a generic set of initiator probes can be used to detect a wide pool of protein targets once a good pair of specific primary antibodies is available. This proposed direction is for developing the split proximity circuit as an actual end-user product, akin to that of the proximity ligation assay [12] but with the convenience of one-pot detection and shorter analysis time (approximately 30 min).

It should be noted that the performance of the split proximity circuit depends on the quality of the recognition moiety. We have demonstrated the compatibility of the circuit with conventional recognition molecules including cDNA, aptamers and antibody. Conceptually, aptamer is best suited for circuit applications since it is straightforward to incorporate relevant sequences at the point of oligonucleotide synthesis. However, its practical applicability is limited to biomolecules with well-characterised and reproducible aptamer sequences, for which the aptamer library is still growing. It is worth considering the use of recombinant antibody, e.g. $F(ab')_2$ fragment and single-chain fragment variable (scfv) molecules, to improve the affinity and specificity of recognition for immediate future works [13].

Lastly, the core of our split proximity circuit is the signal transduction module based on the reassembly of c^* b^* trigger. HCR was a useful signal amplification and target visualization tool within the scope of this thesis. We can consider incorporating other downstream modules with functional outputs, e.g. a DNA origami cage loaded with drugs which unlocks itself based on the panel of biomolecules recognized [14].

9.2.2 Self-contained Logic Gates Design on Cell Surfaces

More complex circuit computations can be performed on cell surface markers. Stojanovic et al. [15] reported the design of an autonomous molecular cascade for profiling the subtypes of immune cells, where up to three cell surface markers can be evaluated by AND, OR, NOT gates through a series of strand displacement reactions. As Schaus and Yin [16] pointed out, the recognition of three random antigens is sufficient for identifying a specific cell type. In the context of cell surface receptor clustering, typically the extracellular domains associate as dimers, e.g. receptor protein-tyrosine kinases and cytokine receptors [17]. There are less common cases of trimer formation, e.g. in tumour necrosis factor alpha (TNFα) through higher order clustering [17].

To achieve a wash-free, single-step reaction, we can build on the concept of implementing DNA walker on microparticle surface by Ellington et al. (Scheme 9.2a) [18]. Two simple AND and OR gates can be designed to evaluate this three-entity complex—a hairpin to lock the readout toehold which is released only when the three receptors come into close proximity (AND gate), while the OR gate is designed by repeating one of the initiator strands on two recognition moieties (Scheme 9.2b). This logic gate designs can be scaled up to detect even higher order assemblies in proximity, e.g. by using more hairpins to execute the cascaded AND gates.

9.2.3 Improving the Performance of HCR

HCR is an attractive enzyme-free signal amplification method [19] which unfortunately, has yet to become a mainstream toolbox, such as rolling circle amplification [20] and DNAzymes amongst many others [21]. Our design guidelines introduced in Chap. 5 can overcome the first barrier to entry, i.e. the design of robust hairpin sequences. However, the truncation of HCR due to defected oligonucleotides often limited our circuit performance despite much efforts to boost signal and reduce background (Sect. 7.2.5).

Acknowledging that biological replications occur with much lower error rates, as elaborated in Sect. 6.2.2, it may be worth encoding the hairpin sequences in plasmid for the large-scale production of highly pure hairpin monomers. The enzymatic method reported by Högberg et al. [22] appears promising for producing stoichiometric amount of single-stranded DNA oligonucleotide with error rates as low as 10^{-8} (Fig. 9.1).

We anticipate that using high quality hairpin monomers can translate to much improved HCR performance, as similarly observed for enzymatically-produced catalytic hairpin assembly hairpins (Sect. 6.2.2), strand displacement gates for implementing chemical reaction networks [23] and formation of DNA nanostructures [24].

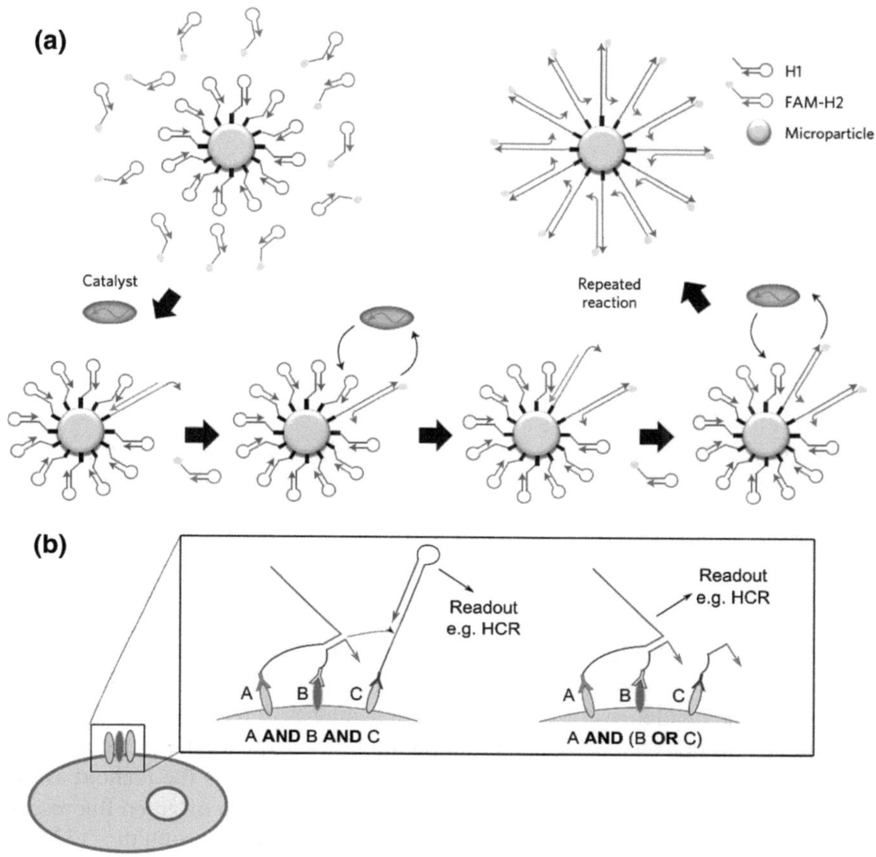

Scheme 9.2 The concept of **a** DNA walker [Reproduced with permission from Ref. [18]. Copyright © 2015, Springer Nature] can be used to **b** design logic gates for probing higher order assemblies

9.3 Future Outlook

Dynamic DNA nanotechnology stands at a critical point where numerous design concepts have been proposed over the past decade, [25] yet the translation of promising concepts into real applications is lagging. The applicability of DNA circuitry is limited by practical considerations, e.g. sensitivity to defected sequences, poor DNA stability and delivery challenges [26, 27]. Our contribution in minimizing circuit leakage by design is just the first yet crucial step for positioning the relevant DNA circuit designs for actual applications.

Moving forward, we can leverage on the concepts in related areas, e.g. structural DNA nanotechnology and synthetic biology, to further improve the circuit performance. First, the combined use with DNA origami should improve the addressability

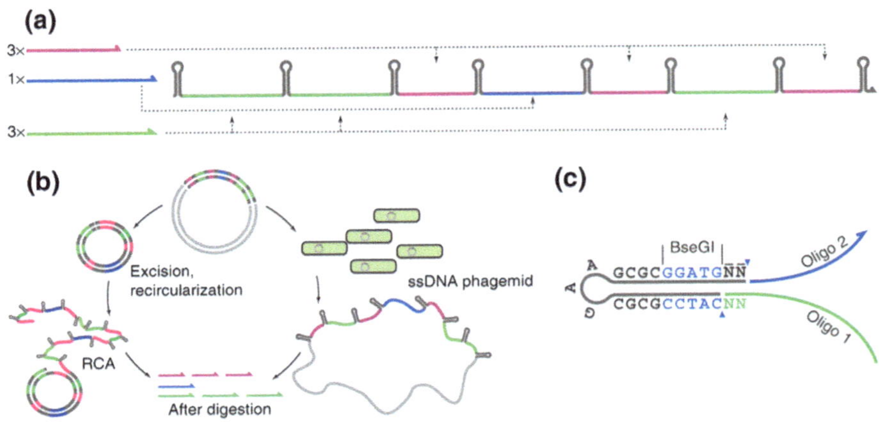

Fig. 9.1 Production of single stranded DNA oligonucleotides by **a** encoding the strands in stoichiometric amount and **b** using phage or enzymes to generate large quantity of the encoded sequences. **c** The individual oligonucleotides are released from the hairpin spacers by digesting with BseGI [Reproduced with permission from Ref. [22]. Copyright © 2013, Springer Nature]

of DNA circuit, which otherwise diffuses randomly in solution, enhance its stability in serum and act as the delivery vehicle into the cells [28–30].

Next, the in vivo expression of DNA circuit design is an interesting direction for overcoming the challenges of oligonucleotide synthesis defect, lack of continuous fuel source and delivery barrier [31]. In addition, the expressed DNA circuit can act as a synthetic regulator of genetic circuits. One example is the toehold switch system where a simple toehold design controls the expression of green fluorescent protein (GFP), [32] which was later incorporated into paper microfluidics [33] and used successfully for Zika virus detection [34].

The ultimate goal is to develop DNA circuits as nanorobots to perform real-time diagnosis and dispense treatment whenever necessary. The recent concept of implementing machine learning in strand displacement network [35] should push the field one step nearer to this end goal once the abovementioned practical limitations are resolved.

References

1. Kiessling, L.L., Gestwicki, J.E., Strong, L.E.: Synthetic multivalent ligands in the exploration of cell-surface interactions. Curr. Opin. Chem. Biol. **4**, 696 (2000)
2. Spengler, M., Adler, M., Niemeyer, C.M.: Highly sensitive ligand-binding assays in pre-clinical and clinical applications: immuno-pcr and other emerging techniques. Analyst **140**, 6175 (2015)
3. Lagerstrom, M.C., Schioth, H.B.: Structural diversity of G protein-coupled receptors and significance for drug discovery. Nat. Rev. Drug Discov. **7**, 339 (2008)

4. Arkin, M.R., Wells, J.A.: Small-molecule inhibitors of protein-protein interactions: progressing towards the dream. Nat. Rev. Drug Discov. **3**, 301 (2004)
5. Vidal, M., Cusick, M.E., Barabási, A.-L.: Interactome networks and human disease. Cell **144**, 986 (2011)
6. Braun, P.: Interactome mapping for analysis of complex phenotypes: insights from benchmarking binary interaction assays. Proteomics **12**, 1499 (2012)
7. Weitsman, G., Barber, P.R., Nguyen, L.K., Lawler, K., Patel, G., Woodman, N., Kelleher, M.T., Pinder, S.E., Rowley, M., Ellis, P.A., Purushotham, A.D., Coolen, A.C., Kholodenko, B.N., Vojnovic, B., Gillett, C., Ng, T.: HER2-HER3 dimer quantification by FLIM-FRET predicts breast cancer metastatic relapse independently of HER2 IHC status. Oncotarget **7**, 51012 (2016)
8. Balic, M., Williams, A., Lin, H., Datar, R., Cote, R.J.: Circulating tumor cells: from bench to bedside. Annu. Rev. Med. **64**, 31 (2013)
9. Crowley, E., Di Nicolantonio, F., Loupakis, F., Bardelli, A.: Liquid biopsy: monitoring cancer-genetics in the blood. Nat. Rev. Clin. Oncol. **10**, 472 (2013)
10. Karkare, S., Bhatnagar, D.: Promising nucleic acid analogs and mimics: characteristic features and applications of PNA, LNA, and morpholino. Appl. Microbiol. Biotechnol. **71**, 575 (2006)
11. Summerton, J.E.: Peptide Nucleic Acids, Morpholinos and Related Antisense Biomolecules, p.89. Springer, Boston, MA, USA (2006)
12. Soderberg, O., Gullberg, M., Jarvius, M., Ridderstrale, K., Leuchowius, K.-J., Jarvius, J., Wester, K., Hydbring, P., Bahram, F., Larsson, L.-G., Landegren, U.: Direct observation of individual endogenous protein complexes in situ by proximity ligation. Nat. Methods **3**, 995 (2006)
13. Hoogenboom, H.R.: Selecting and screening recombinant antibody libraries. Nat. Biotechnol. **23**, 1105 (2005)
14. Douglas, S.M., Bachelet, I., Church, G.M.: A Logic-gated nanorobot for targeted transport of molecular payloads. Science **335**, 831 (2012)
15. Rudchenko, M., Taylor, S., Pallavi, P., Dechkovskaia, A., Khan, S., Butler Jr., V.P., Rudchenko, S., Stojanovic, M.N.: Autonomous molecular cascades for evaluation of cell surfaces. Nat. Nanotechnol. **8**, 580 (2013)
16. Schaus, T.E., Yin, P.: Molecular computing. in situ computation of cell identity. Nat. Nanotechnol. **8**, 546 (2013)
17. Cooper, G.M.: In The Cell: a Molecular Approach, 2nd edn. Sinauer Associates, Sunderland (MA) (2000)
18. Jung, C., Allen, P.B., Ellington, A.D.A.: Stochastic DNA walker that traverses a microparticle surface. Nat. Nanotechnol. **11**, 157 (2016)
19. Ikbal, J., Lim, G.S., Gao, Z.: The hybridization chain reaction in the development of ultrasensitive nucleic acid assays. TrAC Trends Anal. Chem. **64**, 86 (2015)
20. Ali, M.M., Li, F., Zhang, Z., Zhang, K., Kang, D.-K., Ankrum, J.A., Le, X.C., Zhao, W.: Rolling circle amplification: a versatile tool for chemical biology, materials science and medicine. Chem. Soc. Rev. **43**, 3324 (2014)
21. Yan, L., Zhou, J., Zheng, Y., Gamson, A.S., Roembke, B.T., Nakayama, S., Sintim, H.O.: Isothermal amplified detection of DNA and RNA. Mol. BioSyst. **10**, 970 (2014)
22. Ducani, C., Kaul, C., Moche, M., Shih, W.M., Hogberg, B.: Enzymatic production of 'monoclonal stoichiometric' single-stranded dna oligonucleotides. Nat. Methods **10**, 647 (2013)
23. Chen, Y.-J., Rao, S.D., Seelig, G.: Plasmid-derived DNA Strand displacement gates for implementing chemical reaction networks. JoVE. e53087 (2015)
24. Elbaz, J., Yin, P., Voigt, C.A.: Genetic encoding of DNA nanostructures and their self-assembly in living bacteria. Nat. Commun. **7**, 11179 (2016)
25. Zhang, D.Y., Seelig, G.: Dynamic DNA nanotechnology using strand-displacement reactions. Nat. Chem. **3**, 103 (2011)
26. Pinheiro, A.V., Han, D., Shih, W.M., Yan, H.: Challenges and opportunities for structural DNA nanotechnology. Nat. Nanotechnol. **6**, 763 (2011)
27. Ezziane, Z.: DNA computing: applications and challenges. Nanotechnology **17**, R27 (2006)

28. Veneziano, R., Ratanalert, S., Zhang, K., Zhang, F., Yan, H., Chiu, W., Bathe, M.: Designer nanoscale DNA assemblies programmed from the top down. Science (2016)
29. Hahn, J., Wickham, S.F.J., Shih, W.M., Perrault, S.D.: Addressing the instability of DNA Nanostructures in tissue culture. ACS Nano **8**, 8765 (2014)
30. Amir, Y., Ben-Ishay, E., Levner, D., Ittah, S., Abu-Horowitz, A., Bachelet, I.: Universal computing by DNA origami robots in a living animal. Nat. Nanotechnol. **9**, 353 (2014)
31. Chen, Y.-J., Groves, B., Muscat, R.A., Seelig, G.: DNA nanotechnology from the test tube to the cell. Nat. Nanotechnol. **10**, 748 (2015)
32. Green, A.A., Silver, P.A., Collins, J.J., Yin, P.: Toehold switches: De-Novo-designed regulators of gene expression. Cell. **159**, 925 (2014)
33. Pardee, K., Green, A.A., Ferrante, T., Cameron, D.E., DaleyKeyser, A., Yin, P., Collins, J.J.: Based synthetic gene networks. Cell **159**(4), 940–954 (2014)
34. Pardee, K., Green, A.A., Takahashi, M.K., Braff, D., Lambert, G., Lee, J.W., Ferrante, T., Ma, D., Donghia, N., Fan, M., Daringer, N.M.: Rapid, low-cost detection of Zika virus using programmable biomolecular components. Cell **165**(5), 1255–1266 (2016)
35. Lakin, M.R., Stefanovic, D.: Supervised learning in adaptive DNA strand displacement networks. ACS Synth. Biol. **5**, 885 (2016)

Appendix A
List of DNA Sequences Used

All DNA sequences were designed using NUPACK web server to minimise secondary structure and misfoldings [1]. The domains were colour-coded as indicated in the respective chapters (Tables A.1, A.2, A.3 and A.4).

A. Yan Shan, *Engineering a Robust DNA Circuit for the Direct Detection of Biomolecular Interactions*, Springer Theses,
https://doi.org/10.1007/978-981-13-2188-7

Table A.1 List of DNA sequences used to design a self-contained DNA circuit (Chap. 4)

Strand	Name	Sequence
Optimization of Molecular Switch Design		
Design i	-	GGT TGG TGT GGT TGG ACA CCA ACC AGT CTA GGA TTC GGC GTG
Design ii	-	GGT TGG TGT GGT TGG CAA CCA GTC TAG GAT TCG GCG TG
Design iii	-	CTG GTT GGT GTG GTT GGC AAC CAG TCT AGG ATT CGG CGT G
Design iv	-	GAC TGG TTG GTG TGG TTG GCA ACC AGT CTA GGA TTC GGC GTG
Design v	-	CAA CCA GTC TAG GAT TCG GCG TGC TAG ACT GGT TGG TGT GGT TGG
Module Kinetics – Hybridization Chain Reaction		
Hairpin 1	**HP1**	TTA ACC CAC GCC GAA TCC TAG ACT CAA AGT CTA GGA TTC GGC GTG
Hairpin 2	**HP2**	AGT CTA GGA TTC GGC GTG GGT TAA CAC GCC GAA TCC TAG ACT ACT TTG
c* b*	-	AGT CTA GGA TTC GGC GTG GGT TAA
Module Kinetics – Signal Transduction		
Protector	**P**	TTA ACC CAC GCC GAA TCC TAG ACT GGT TG
t* c*	-	CA ACC AGT CTA GGA TTC GGC GTG
Module Kinetics - α-Thrombin Recognition		
Initiator 1	**I1**	Refer to sequence of molecular switch design iii.
Initiator 2	**I2**	AGT CTA GGA TTC GGC GTG GGT TAA TTT TTT TTT TTT TTT AGT CCG TGG TAG GGC AGG TTG GGG TGA CT
Formation of Localized Signal		
I2-15 thymine	**I2-15T**	Refer to sequence of I2.
I2-5 thymine	**I2-5T**	AGT CTA GGA TTC GGC GTG GGT TAA TTT TTA GTC CGT GGT AGG GCA GGT TGG GGT GAC T
Unbound	**noApt**	AGT CTA GGA TTC GGC GTG GGT TAA

Table A.2 List of DNA sequences used to generate design guidelines for hybridization chain reaction (HCR) hairpins (Chap. 5)

Strand	Name	Sequence
Experimental Validation of Design Guidelines		
Pierce original hairpin 1	-	TTA ACC CAC GCC GAA TCC TAG ACT CAA AGT AGT CTA GGA TTC GGC GTG
Pierce original hairpin 2	-	AGT CTA GGA TTC GGC GTG GGT TAA CAC GCC GAA TCC TAG ACT ACT TTG
Pierce original trigger	-	AGT CTA GGA TTC GGC GTG GGT TAA
Pierce modified hairpin 1	-	TTA ACC CAC GCC GAA TCC CAA AGT GGA TTC GGC GTG
Pierce modified hairpin 2	-	GGA TTC GGC GTG GGT TAA CAC GCC GAA TCC ACT TTG
Pierce modified trigger	-	GGA TTC GGC GTG GGT TAA
New hairpin 1	**HP1**	GGA ATT GGG AGT AAG GGC TGT GAT GCC CTT ACT CCC
New hairpin 2	**HP2**	GCC CTT ACT CCC AAT TCC GGG AGT AAG GGC ATC ACA
New trigger	**T**	GCC CTT ACT CCC AAT TCC
Quencher for Kinetics Study		
Protector 1	**P1**	GGA ATT GGG AGT AAG GGC
Protector 2	**P2**	GGG AGT AAG GGC ATC ACA
HCR Förster Resonance Energy Transfer (FRET) Design		
New hairpin 1 tagged with 5'-Cy3	**HP1-Cy3**	/5-Cy3/ GGA ATT GGG AGT AAG GGC TGT GAT GCC CTT ACT CCC
New hairpin 2 tagged with 3'-Cy5	**HP2-Cy5**	GCC CTT ACT CCC AAT TCC GGG AGT AAG GGC ATC ACA /3-Cy5/
Fluorophore strand	**F**	GTT GGA ATT GGG AGT AAG GGC /3-FAM/
Quencher strand	**Q**	/5-IABkFQ/ GCC CTT ACT CCC
HCR Labelling Schemes		
HP1 tagged with internal Cy5	**IntCy5**	GTT GGA ATT GGG AGT AAG GGC /iCy5/ TGT GAT GCC CTT ACT CCC
HP1-iCy5 with 1T spacer	**IntCy5 (+1T)**	GTT GGA ATT GGG AGT AAG GGC **T** /iCy5/ TGT GAT GCC CTT ACT CCC
HP2 with end Cy3	**HP2_Cy3**	GCC CTT ACT CCC AAT TCC AAC GGG AGT AAG GGC ATC ACA /3-Cy3/

Table A.3 List of DNA sequences used to study leakages involved in split proximity circuit (Chap. 6)

Strand	Name	Sequence
Trigger Strand		
Direct Trigger	**DT**	GCC CTT ACT CCC AAT TCC AAC
Hairpin Trigger	**HT**	GCC CTT ACT CCC GCG GCG CGC CTT GGC GCG CCG CTT AAT TCC AAC
Split Trigger	**ST**	GAG TGG ATG GTG AAG GTG AAG GTA
Reporter Signals		
Fluorescein-tagged readout	**F**	GTT GGA ATT GGG AGT AAG GGC /3-FAM/
Dabcyl-Tagged Quencher	**Q**	/5-Q/ GCC CTT ACT CCC
Hairpin 1	**HP1**	GTT GGA ATT GGG AGT AAG GGC TGT GAT GCC CTT ACT CCC
Hairpin 2	**HP2**	GCC CTT ACT CCC AAT TCC AAC GGG AGT AAG GGC ATC ACA
Association Concept		
Initiator 1(no association)	-	TAC CTT CAC CTT TTT TTT TTT TTT TTT TT AAT TCC AAC
Initiator 2(no association)	-	GCC CTT ACT CCC TTT TTT TTT TTT TTT CAC CAT CCA CTC
Leak I and II		
Initiator 2-Design 1	**I2**	GCC CTT ACT CCC GGT CAC TTT TTT TTT TTT TTT CAC CAT CCA CTC
I2 with 3' defect	**I2 (-3')**	GCC CTT ACT CC_ GGT CAC TTT TTT TTT TTT TTT CAC CAT CCA CTC
I2 with 5' defect	**I2 (-5')**	_CC CTT ACT CCC GGT CAC TTT TTT TTT TTT TTT CAC CAT CCA CTC
Initiator 1 – 9 nt domain *b*	-	TAC CTT CAC CTT TTT TTT TTT TTT TTT GTG ACC TTA ATT CCA AC
Initiator 2-Design 2-Lock	**I2-D2L**	GGG AGT AAG GGC TTT
Initiator 2-Design 3	**I2-D3**	GCC CTT ACT CCG GGT CAC TTT TTT TTT TTT TTT CAC CAT CCA CTC
Initiator 2-Design 4	**I2-D4**	GCC CTT ACT CC_ GGT CAC TTT TTT TTT TTT TTT CAC CAT CCA CTC
Initiator 1 - 10nt domain *b*	**I1**	TAC CTT CAC CTT TTT TTT TTT TTT TTT GTG ACC TTC AAT TCC AAC

Study of Intermediate Complexes Responsible for Background Leakage

Initiator 1 (I1-*length of domain a-length of domain b*)	**I1-a3-b9**	TAC CTT CAC CTT TTT TTT TTT TTT TTT GTG TT C AAT TCC AAC
	I1-a4-b9	TAC CTT CAC CTT TTT TTT TTT TTT TTT GTG C TT C AAT TCC AAC
	I1-a4-b6	TAC CTT CAC CTT TTT TTT TTT TTT TTT GTG C TT C AAT TCC
	I1-a5-b9	TAC CTT CAC CTT TTT TTT TTT TTT TTT GTG AC TT C AAT TCC AAC
	I1-a6-b9	Refer to I1-b10
	I1-a6-b7	TAC CTT CAC CTT TTT TTT TTT TTT TTT GTG ACC TT C AAT TCC A
	I1-a6-b5	TAC CTT CAC CTT TTT TTT TTT TTT TTT GTG ACC TT C AAT TC
	I1-a6-b4	TAC CTT CAC CTT TTT TTT TTT TTT TTT GTG ACC TT C AAT T
Initiator 2 (I2-*length of domain a*)	**I2-a3**	GCC CTT ACT CC CAC TTT TTT TTT TTT TTT CAC CAT CCA CTC
	I2-a4	GCC CTT ACT CC G CAC TTT TTT TTT TTT TTT CAC CAT CCA CTC
	I2-a5	GCC CTT ACT CC GT CAC TTT TTT TTT TTT TTT CAC CAT CCA CTC
	I2-a6	Refer to I2-D4

Streptavidin-Biotin as Model System

Streptavidin-Initiator 1	**Stav-I1**	/5-Biotin/TTT TTT TTT TTT TTT GTG CTT CAA TTC CAA C
Streptavidin-Initiator 2	**Stav-I2**	GCC CTT ACT CCG CAC TTT TTT TTT TTT TTT /3-Biotin/

Confirming Leakage Observation with New Sequence Set

Fluorescein-tagged readout	**F2**	CAC CTT AAA CCT CAC CCT ACC /3-FAM/
Dabcyl-Tagged Quencher	**Q2**	/5-Q/ GGT AGG GTG AGG
Direct trigger	**DT2**	GGT AGG GTG AGG TTT AAG GTG
Initiator 1	**I1(2)**	TGT TTA TGG AGT TTT TTT TTT TTT TTT CTCG TT TTT AAG GTG
Initiator 2	**I2(2)**	GGT AGG GTG AGG CGAG TTT TTT TTT TTT TTT GTA GGA GTG GAT

I2 with 5' defect	**I2(2,-5')**	_GT AGG GTG AGG CGAG TTT TTT TTT TTT TTT GTA GGA GTG GAT
I1 (with translocation)	**I1'(2)**	TGT TTA TGG AGT TTT TTT TTT TTT TTT CTCG TT G TTT AAG GTG
I2 (with translocation)	**I2'(2) / I2 (2,-3')**	GGT AGG GTG AG_ CGAG TTT TTT TTT TTT TTT GTA GGA GTG GAT
Split trigger	**ST2**	ATC CAC TCC TAC T ACT CCA TAA ACA

Table A.4 List of DNA sequences used in the final split proximity circuit (Chaps. 7 and 8)

Strand	Name	Sequence
Trigger Strand		
Direct Trigger	**DT**	GCC CTT ACT CCC AAT TCC AAC
Hairpin Trigger	**HT**	GCC CTT ACT CCC GCG GCG CGC CTT GGC GCG CCG CTT AAT TCC AAC
Split Trigger	**ST**	GAG TGG ATG GTG AAG GTG AAG GTA
Reporter Signals		
Hairpin 1 tagged with internal Cy5	**HP1**	GTT GGA ATT GGG AGT AAG GGC /iCy5/TGT GAT GCC CTT ACT CCC
Hairpin 2 tagged with Cy3	**HP2**	GCC CTT ACT CCC AAT TCC AAC GGG AGT AAG GGC ATC ACA /Cy3/
Fluorescein-tagged readout	**F**	GTT GGA ATT GGG AGT AAG GGC/3-FAM/
Dabcyl-Tagged Quencher	**Q**	/5-Q/ GCC CTT ACT CCC
Circuit Design for DNA Study		
Initiator 1 (for split trigger)	**I1**	TAC CTT CAC CTT TTT TTT TTT TTT TTT GTG C TT C AAT TCC AAC
Initiator 2 (for split trigger)	**I2**	GCC CTT ACT CC G CAC TTT TTT TTT TTT TTT CAC CAT CCA CTC
Split trigger (12-0-12)	**ST12-0-12**	GAG TGG ATG GTG AAG GTG AAG GTA
Split trigger (12-1-12)	**ST12-1-12**	GAG TGG ATG GTG T AAG GTG AAG GTA
Split trigger (11-0-11)	**ST11-0-11**	AGT GGA TGG TGA AGG TGA AGG T
Split trigger (10-0-10)	**ST10-0-10**	GTG GAT GGT GAA GGT GAA GG
Split trigger (9-0-9)	**ST9-0-9**	TGG ATG GTG AAG GTG AAG
Split trigger (8-0-8)	**ST8-0-8**	GGA TGG TGA AGG TGA A
Split trigger (12-0-9)	**ST12-0-9**	GAG TGG ATG GTG AAG GTG AAG
ST1209 (end G>A)	**SM1**	GAG TGG ATG GTG AAG GTG AA**<u>A</u>**
ST1209 (18G>A)	**SM2**	GAG TGG ATG GTG AAG GT**<u>A</u>** AAG
ST1209 (13A>T)	**SM3**	GAG TGG ATG GTG **<u>T</u>**AG GTG AAG
Anti-split trigger	**aST**	TAC CTT CAC CTT CAC CAT CCA CTC

Circuit Design for miRNA Detection		
I1 (for let7 miRNA)	-	AAC TAT ACA AC TTT TTT TTT TTT TTT GTG C TT C AAT TCC AAC
I2 (for let7 miRNA)	-	GCC CTT ACT CCG CAC TTT TTT TTT TTT TTT TAC TAC CTC A
let7a miRNA	**let7a**	UGA GGU AGU AGG UUG UAU AGU U
let7b miRNA	**let7b**	UGA GGU AGU AGG UUG U**<u>G</u>**U **<u>G</u>**GU U
let7c miRNA	**let7c**	UGA GGU AGU AGG UUG UAU **<u>G</u>**GU U
let7d miRNA	**let7d**	<u>A</u>GA GGU AGU AGG UUG **<u>C</u>**AU AGU U
Circuit Design for Thrombin Detection		
Hairpin 1 tagged with internal Cy5	**HP1**	CAC CTT AAA CCT CAC CCT ACC /iCy5/ TCA AGT GGT AGG GTG AGG
Hairpin 2 tagged with Cy3	**HP2**	GGT AGG GTG AGG TTT AAG GTG CCT CAC CCT ACC ACT TGA /3-Cy3/
I1 (heparin recognition)	**I1-apt29**	AGT CCG TGG TAG GGC AGG TTG GGG TGA CTT TTT TTT TTT TTT TTC TCG TTG TTT AAG GTG
I2 (fibrinogen recognition)	**I2-apt15**	GGT AGG GTG AGC GAG TTT TTT TTT TTT TTT TTT GGT TGG TGT GGT TGG
Aptamer for heparin recognition site	**apt29_hep**	AGT CCG TGG TAG GGC AGG TTG GGG TGA CT
Aptamer for fibrinogen recognition site	**apt15_fib**	GGT TGG TGT GGT TGG
Directly modified I1-apt15	-	GGT TGG TGT GGT TGG TTT TTT TTT TTT TTT GTG CTT CAA TTC CAA C
Directly modified I2-apt29	-	GCC CTT ACT CCG CAC TTT TTT TTT TTT TTT AGT CCG TGG TAG GGC AGG TTG GGG TGA CT
General Circuit Design for Protein Detection		
DT tagged with biotin	**DT**	GCC CTT ACT CCC AAT TCC AAC TTT TTT TTT TTT TTT /3Bio/
I1 tagged with biotin	**I1**	/5Biosg/TTT TTT TTT TTT TTT GTG CTT CAA TTC CAA C
I2 tagged with biotin	**I2**	GCC CTT ACT CCG CAC TTT TTT TTT TTT TTT /3Bio/

Appendix B
Example of NUPACK Design Code

Before setting up the NUPACK code, one should have thought of the desired structure (duplex or single-stranded) and domain design (length and sequence constraints, if any).

```
# Set up system parameters
#
Material = dna
temperature[C] = 25.0
trials = 10
sodium[M] = 0.48
dangles = some

#
# Define secondary structure using DU+ notation
# U = unpaired; D = duplex
#
structure HP1 = U9 D12 U6
structure HP2 = D12 U9 U6
structure I1 = U42
structure I2 = U43
structure ST = U24

#
# Define the sequence domains (length and sequence constraints)
# N = nucleotide
#
domain a = N4
domain b = GTT GGA ATT
domain c = GGG AGT AAG GGC
domain d = TGT GAT
```


A. Yan Shan, *Engineering a Robust DNA Circuit for the Direct Detection of Biomolecular Interactions*, Springer Theses,
https://doi.org/10.1007/978-981-13-2188-7

```
domain s = TTT TTT TTT TTT TTT
domain t = TT
domain x = N12
domain y = N12

#
# Thread the sequence domains onto desired structures
# * = complementary domain; thread in 5' to 3' direction
#
HP1.seq = b c d c*
HP2.seq = c* b* c d*
I1.seq = x s a t b*
I2.seq = c* a* s y
ST.seq = y* x*

#
# Specify stop conditions for normalized ensemble defect
# Default = 1.0 (percent) for each structure
#
HP1.stop = 0.1
HP2.stop = 0.1
I1.stop = 0.1
I2.stop = 0.1
ST.stop = 0.1

#
# Prevent the following sequence patterns
# Constraint on K, Y and W not implemented as polyT is inherent in design
prevent = AAAA, CCCC, GGGG, MMMMMM, RRRRRR, SSSSSS
```

Appendix C
Sulfo-SMCC Method for Antibody-DNA Conjugation

Sulfo-SMCC is a heterobifunctional crosslinker that can covalently conjugate amine-containing antibody and thiolated DNA. Briefly, the antibody solution was first washed with spin column to remove amine-containing azide, a common additive used to curb bacteria contamination, followed by concentration to at least 1 mg/mL. 4 mM sulfo-SMCC (dissolved in anhydrous DMSO) was added in 100:1 molar excess in protein activation buffer (55 mM phosphate buffer, 150 mM NaCl, 5 mM EDTA, pH 7.2). The mixture was incubated at RT for 2 h. Gel filtration was used to purify the activated antibody with a recovery of ca. 70%.

Simultaneously, 6:1 molar excess (found to be optimal loading ratio) of thiolated DNA to antibody was prepared by cleaving the thiolated group with 100:1 molar excess of dithiothreitol in DNA activation buffer (55 mM phosphate buffer, 150 mM NaCl, 20 mM EDTA, pH 7.2) for 1 h at 37 °C. The DNA was purified using DNA ethanol precipitation by adding 1/10 volume of sodium acetate, 0.01 mM $MgCl_2$ (final concentration) and 2.5X volume excess of pure ethanol, and incubating at 4 °C for 30 min. The DNA pellet was collected by spinning down at 14k rpm for 15 min, and the salt contaminant was removed by washing with 70% ethanol. The recovery was 90% for this step.

Subsequently, both DNA and antibody were dispersed in or buffer-exchanged to the conjugation buffer (1X PBS, 5.0 mM EDTA, pH 7.4) and mixed together. They were mixed together directly and incubated overnight at 4 °C though the reaction should be completed within 2 h. Excess DNA was then removed using ultra-filtration.

The antibody-DNA recovery was ca. 40% relative to the input antibody amount for this conjugation step. The overall yield in terms of the amount of DNA successfully conjugated at this step was ca. 30%. The formation of antibody-DNA conjugates was confirmed using PAGE analysis (Fig. C.1). While the protocol

A. Yan Shan, *Engineering a Robust DNA Circuit for the Direct Detection of Biomolecular Interactions*, Springer Theses,
https://doi.org/10.1007/978-981-13-2188-7

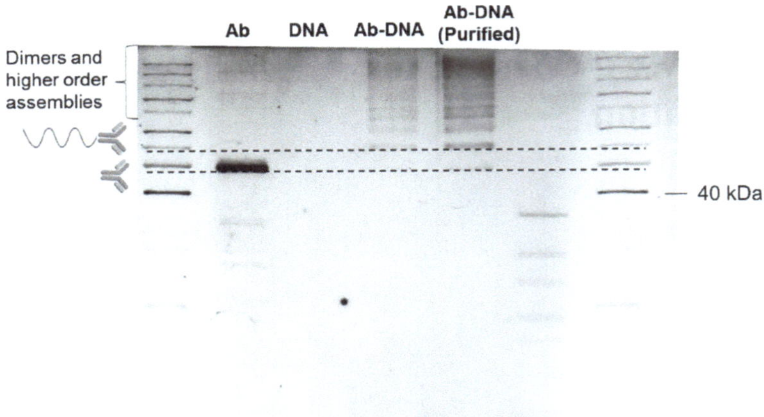

Fig. C.1 PAGE image of the antibody (Ab)-DNA conjugates obtained using sulfo-SMCC conjugation

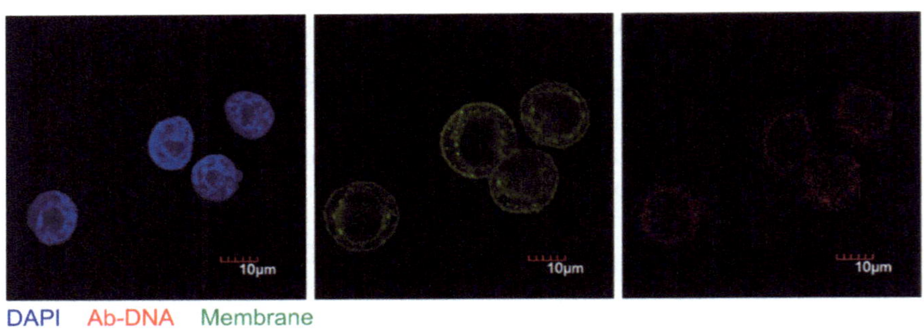

Fig. C.2 Confocal imaging to evaluate the binding of antibody-DNA conjugate (red) to AU565 cells

appeared to be successful from a chemistry point of view, it required further optimization as the bio-recognition ability of the antibody after conjugated appeared to be compromised (Fig. C.2).

No appreciable binding to the cell membrane, where the target HER2 was located, was detected. This image was provided by Dr Chua Pei Jou from Prof Bay's lab in the NUS Department of Anatomy.

Appendix D
NUPACK Simulation for Evolution of HCR Hairpins

The following simulation work was performed by Ben Spanos from the University College of London under my direct mentorship for the UCL CMMP Summer Project. The essence of our simulation work is summarized below.

In all the simulations below, leakage was defined as all HCR complexes that formed in absence of **DT**, e.g. **HP1-HP2**, **HP1-HP2-HP1**, for a three-strand complex system. The input hairpin concentrations were kept constant as 500 nM; while the salt condition of 5X SSCT, i.e. [Na^+] = 0.75 M and [Mg^{2+}] = 0 M was used.

We took the domain configuration of Pierce's original hairpin design (stem length = 18 nt, stem CG% = 55%, toehold CG% = 33%) as the starting point. The toehold length was varied between 6–12 nt. The upper limit of 12 nt is recommended by Pierce's group (Guideline #2) [2]. No leakage was observed. The toehold CG% was then varied for two toehold lengths of 10 and 12 nt. Higher

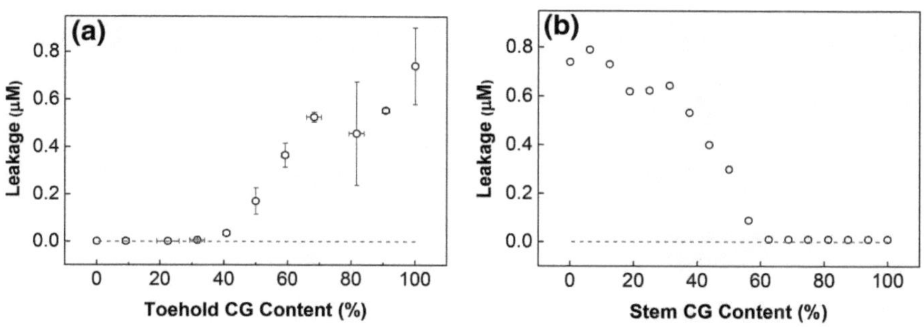

Fig. D.1 HCR leakage computed from NUPACK simulation **a** when the toehold CG content was varied for 25 hairpin sequences and **b** when the stem CG content was varied for 15 sequences for a stem length of 18 nt. Data are shown as mean ± S.D

A. Yan Shan, *Engineering a Robust DNA Circuit for the Direct Detection of Biomolecular Interactions*, Springer Theses,
https://doi.org/10.1007/978-981-13-2188-7

toehold CG% promoted more spurious hybridization and hence leakages. The breakthrough point was about 30–40% toehold CG (Fig. D.1a) (Guideline #1).

The toehold design was then kept constant (length = 10 nt, CG% = 40%) and the stem CG% was varied while keeping the stem length constant as 18 nt. The higher the stem CG%, the more stable the hairpins were in absence of **DT**. 60% CG content was effective in supressing the leakages (Fig. D.1b).

Other minor points were noted from the simulation works, though they only improved the generation of HCR signal marginally ("more or less" outcome) rather than having a drastic effect as in leakages ("yes or no" outcome). First, the HCR signal was maximized when the length of the toehold was longer than the loop in **HP1**. This was due to the unequal role played by the two hairpins where **HP1** participated in hybridizing with both **DT** (initiation) as well as **HP2** (propagation). As such, a longer toehold length for **HP1** favoured initiation over propagation for more efficient use of available **DT** strands. For such asymmetric toehold/loop lengths, we found that the Guideline #1 was still valid, i.e. the C•G% of both toehold/loop must be less than 30–40%. Next, it was favourable to place the C•G pairs at the edge of the toehold (away from the stem) while there was minimal positional effect for the stem.

Appendix E
Förster Resonance Energy Transfer Calculations

Förster resonance energy transfer (FRET) refers to the nonradiative energy transfer from an excited donor molecule (Cy3) to an acceptor molecule (Cy5) due to their spectral overlap (Fig. E.1) [3]. It is described by the Förster radius (R_o) which is the center-to-center separation distance (r) where the energy transfer efficiency is 50%. This parameter was calculated using Eq. (E.1).

$$R_o = (B\,\Phi_D I)^{1/6} \quad (E.1)$$

where,

B and Iwere calculated using Eqs. (E.2) and (E.3) respectively.

Φ_D = quantum yield of **HP1-Cy3** (determined to be 0.277 using rhodamine 6G as reference dye).

$$B = \frac{9000[\ln(10)]\kappa_p^2}{128\pi^5 n_D^4 N_A} \quad (E.2)$$

where,

κ_p^2 = dipole orientation factor (range from 0–4; in this case, κ = 2/3 for the randomly oriented dipole);

n_D = refractive index (1.33 for water);

N_A = Avogadro's number

The overlap integral, I, measures the extent of spectral overlap between the donor and acceptor dipole and is described by Eq. (E.3) and plotted in Fig. E.2a.

$$I = \int_0^\infty PL_{D-corr}(\lambda)\ \epsilon_A\ (\lambda)\lambda^4 d\lambda \quad (E.3)$$

A. Yan Shan, *Engineering a Robust DNA Circuit for the Direct Detection of Biomolecular Interactions*, Springer Theses,
https://doi.org/10.1007/978-981-13-2188-7

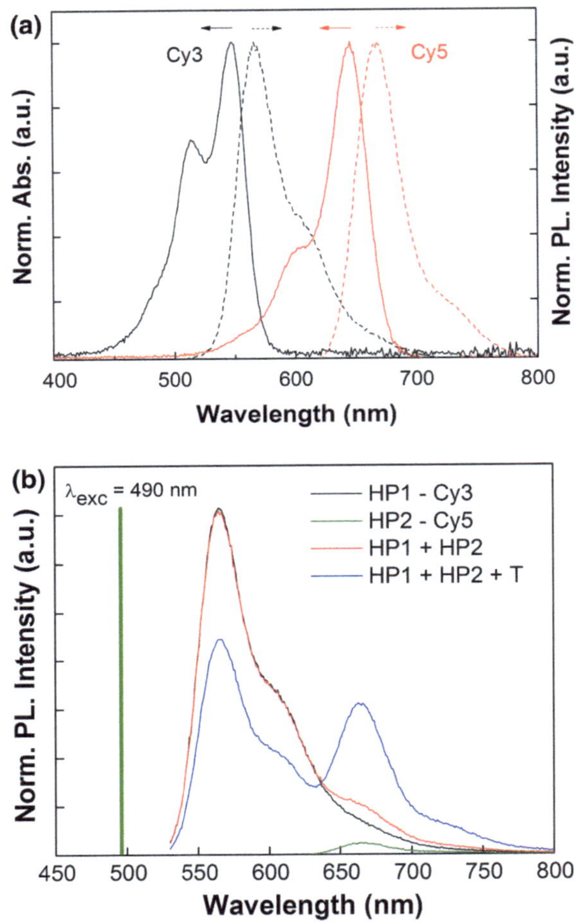

Fig. E.1 Photophysical properties of Cy3 and Cy5 for use in HCR FRET. **a** Normalized absorbance (solid line) and emission (dotted line) spectrum of HP1-Cy3 (black) and HP2-Cy5 (red). Energy transfer took place within the overlap region (grey). **b** At the excitation wavelength (λ_{exc}) of 490 nm, there was minimal emission from Cy5-HP2 (green line) while Cy3-HP1 (black line) generated high fluorescence emission. A mixture of Cy3-HP1 and Cy5-HP2 resulted in obvious FRET in the presence of the trigger strand (blue line) as compared to the significantly lower background noise in absence of the trigger strand (red line)

Where,
ϵ_A = acceptor absorption spectrum;
$PL_{D\text{-}corr}$ = normalized donor emission spectrum.
Solving for I give $8.31 \times 10^{15}\ M^{-1}\ cm^{-1}\ nm^4$.
Substituting back to Eq. (E.1) gives $\mathbf{R_o = 56.9}$ **Å**.
The average energy transfer efficiency (E) was calculated using Eq. (E.4).

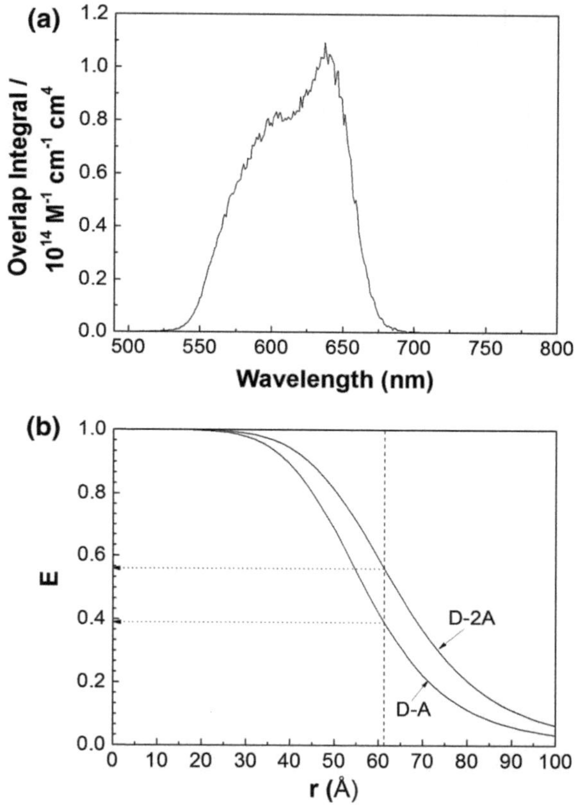

Fig. E.2 Calculation of FRET efficiency. **a** The overlap integral curve was used to calculate the Förster distance (R) as 56.9 Å. **b** The simulated FRET efficiency (E) was plotted for both single donor-acceptor pair (D-A) and donor-double acceptor pair (D-2A). The experimental E obtained was 0.382 for a fixed separation distance (r) of ca. 61.2 Å (dotted line) which can be better explained by a single donor-acceptor pair (D-A) than a two acceptors model (D-2A)

$$E = \frac{(F_D - F_{DA})}{F_D} \tag{E.4}$$

where F_D and F_{DA} are the fluorescence intensities of Cy3 only (donor) and Cy3 in presence of Cy5 (acceptor).

E can also be calculated theoretically based on the Förster dipole-dipole formalism expressed in Eq. (E.5).

$$E = \frac{n\,R_0^6}{n\,R_0^6 + r^6} \tag{E.5}$$

where n is the average number of Cy5 acceptor per Cy3 donor molecule.

In our HCR FRET system, each Cy3 molecule has two Cy5 neighbours along the one-dimensional DNA assembly, suggesting that n = 2(D-2A). At the same time, each Cy5 molecule has two Cy3 neighbours as well for the energy transfer process and possibly possesses only 50% probability of accepting the energy from either Cy3 neighbour. In this case, n = 1(D-A). The simulation results of the two scenarios are shown in Fig. E.2b. For a fixed separation distance of ca. 61.2 Å, the single donor-acceptor energy transfer model can better describe the maximum E value of 0.382 obtained experimentally.

Appendix F
Understanding the Individuals Steps of HCR

Hybridization chain reaction (HCR) can be thought of as a polymerization reaction using DNA as monomers. The overall reaction can be broken down into three key steps (Scheme F.1). First, the trigger strand (**T**) binds to hairpin 1 (**HP1**) to initiate HCR. Once **HP1** opens, the cascaded hairpin opening becomes a sustained process where **HP1** and **HP2** fuel the chain propagation alternately. Finally, HCR terminates either when all hairpin monomers are exhausted, or when defected hairpins bind to the growing chain [4].

The relative extent of HCR initiation and propagation is controlled by the length of domain *b* and *d* (Fig. F.1). For example, longer domain *b* over domain *d* favours initiation and hence the utilization of **T**, while propagation proceeds at a relatively slower rate. On the other hand, a longer domain *d* favours propagation and hence the development of longer DNA chain per **T** strand. A balance between the two can

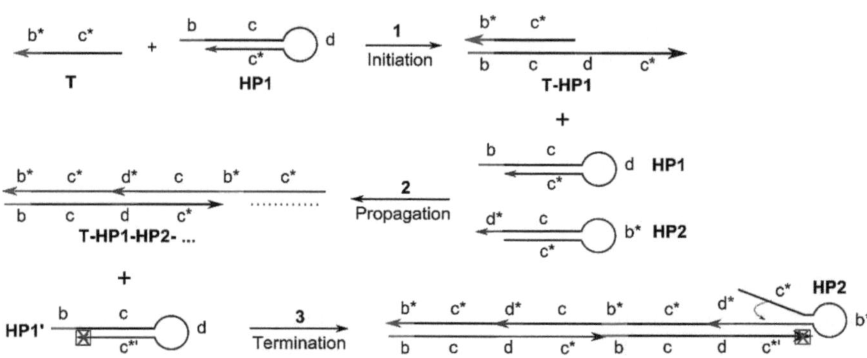

Scheme F.1 HCR can be broken down into three key steps of initiation, propagation and termination

A. Yan Shan, *Engineering a Robust DNA Circuit for the Direct Detection of Biomolecular Interactions*, Springer Theses,
https://doi.org/10.1007/978-981-13-2188-7

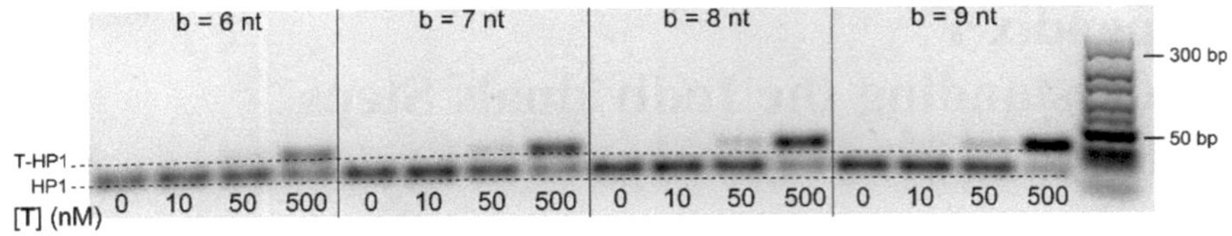

Fig. F.1 Extent of HCR initiation was dependent on the toehold length (domain *b*) 500 nM of hairpin 1 (**HP1**) was reacted with the trigger (**T**) concentations indicated below the gel at 25 °C for 1 h. A longer toehold was more efficient in utilizing the trigger to form more **T-HP1** complex. 10–300 bp DNA ladder is shown on the right of the gel

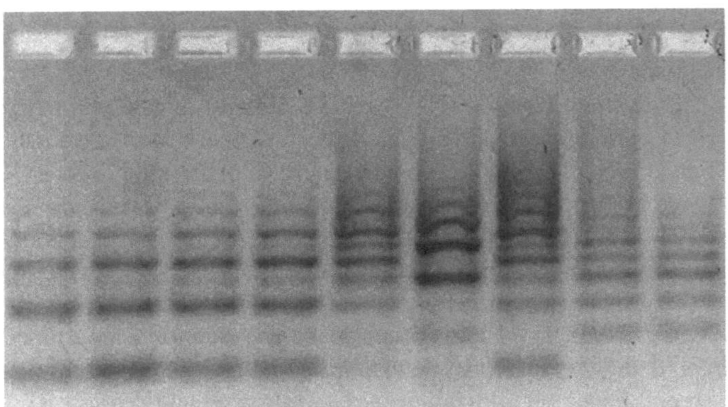

[HP1] (nM)	250	300	400	500	750	1000	1000	500	500
[HP2] (nM)	500	500	500	500	500	500	1000	400	250

Fig. F.2 Gel image of HCR products formed when varying **HP1**-to-**HP2** ratio was used 500 nM of **T** was reacted with the respective **HP1** and **HP2** concentration indicated below the gel at 25 °C for 1 h. This gel image was taken using a digicam under UV lamp

be optimized to control the HCR progression depending on the specific application. We prioritized the efficient utilization of targets available over the signal gain per target, and a configuration of domain $b = 9$ nt and domain $d = 6$ nt was used for the application works shown in Chaps. 7 and 8.

Another factor to note is the ratio of **HP1** to **HP2**. Since **HP1** participates in both HCR initiation and propagation, it should be added in excess amount so that it does not limit the reaction (Fig. F.2). HCR proceeded to similar extent when **HP1** was less than **HP2**, regardless of the actual reactant ratio used. Hairpins were utilized most efficiently when the **HP1-**to-**HP2** ratio was between 1:1 and 2:1.

Appendix G
Characterisation of Cell Lines and Antibodies

Three cell lines were used to evaluate the performance of our split proximity circuit in detecting HER2:HER2 and HER2:HER3 dimers in cell lysates (Sect. 7.2.7) and directly on cell membranes (Chap. 8). The cell lines involved were: AU565 with high expression of HER2 and HER3 [5], MDA-MB-231 which is a triple negative breast cancer cell line [6] and MCF10A which is a normal breast epithelial cell line [7]. The HER2 and HER3 expression levels were confirmed using western blot (Fig. G.1). The high expression levels in AU565 were predominantly observed in the membrane fraction while the much lower expression level in the cytosolic fraction might be artefacts from the two-step protein extraction procedure. There were no visible HER2 and HER3 bands in the MDA-MB-231 and MCF10A samples.

We confirmed the binding affinity of the anti-HER2 affibody and anti-HER3 antibody by titrating increasing concentrations of the antibodies on the fixed AU565 cells (Fig. G.2).

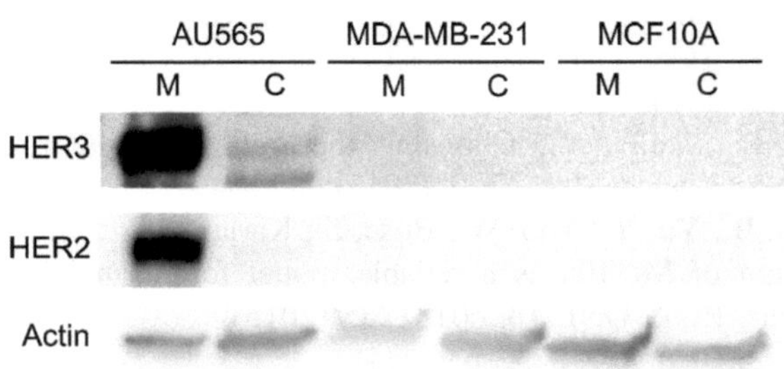

Fig. G.1 Expression levels of HER2 and HER3 proteins in the membrane (M) and cytosolic (C) fraction of AU565, MDA-MB-231 and MCF10A cell lines. Both HER2 and HER3 were highly expressed in AU565, and were of negligible amount in MDA-MB-231 and MCF10A

A. Yan Shan, *Engineering a Robust DNA Circuit for the Direct Detection of Biomolecular Interactions*, Springer Theses,
https://doi.org/10.1007/978-981-13-2188-7

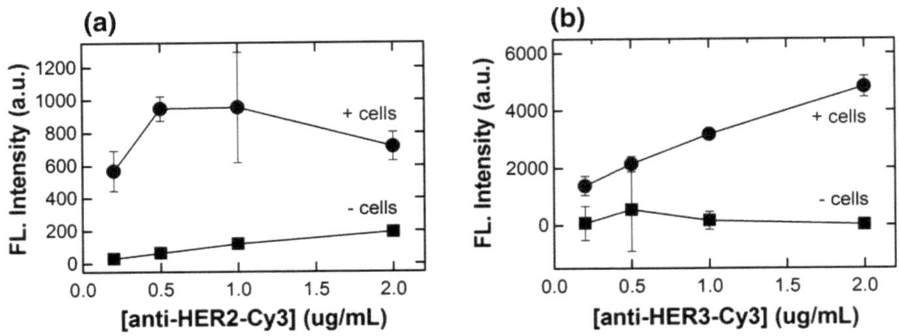

Fig. G.2 The binding affinity of the antibodies was confirmed by titrating increasing concentrations (0.2–2.0 µg/mL) of **a** anti-HER2 and **b** anti-HER3 biotinylated antibody tagged with streptavidin-Cy3

References

1. Zadeh, J.N., Steenberg, C.D., Bois, J.S., Wolfe, B.R., Pierce, M.B., Khan, A.R., Dirks, R.M., Pierce, N.A.: NUPACK: analysis and design of nucleic acid systems. J. Comput. Chem. **32**, 170 (2011)
2. Choi, H.M.T., Beck, V.A., Pierce, N.A.: Next-generation in situ hybridization chain reaction: higher gain, lower cost, greater durability. ACS Nano **8**, 4284 (2014)
3. Clapp, A.R., Medintz, I.L., Mauro, J.M., Fisher, B.R., Bawendi, M.G., Mattoussi, H.: Fluorescence resonance energy transfer between quantum dot donors and dye-labeled protein acceptors. J. Am. Chem. Soc. **126**, 301 (2004)
4. Li, B., Jiang, Y., Chen, X., Ellington, A.D.: Probing spatial organization of DNA strands using enzyme-free hairpin assembly circuits. J. Am. Chem. Soc. **134**, 13918 (2012)
5. Jeong, H., Kim, J., Lee, Y., Seo, J. H., Hong, S. R., Kim, A.: Neuregulin-1 induces cancer stem cell characteristics in breast cancer cell lines. Oncol. Rep. **32**, 1218 (2014)
6. Holliday, D.L., Speirs, V.: Choosing the right cell line for breast cancer research. Breast Cancer Res. **13**, 1 (2011)
7. Qu, Y., Han, B., Yu, Y., Yao, W., Bose, S., Karlan, B.Y., Giuliano, A.E., Cui, X.: Evaluation of Mcf10a as a reliable model for normal human mammary epithelial cells. PLoS ONE **10**, e0131285 (2015)

Appendix H
List of Publications

Publications in Peer-Reviewed Journals/Book Chapters

(*Works included in thesis*)

(1) Ang, Y.S., Li J.-E.J., Chua, P.-J., Ng, C.-T., Bay, B.-H. and Yung, L.-Y.L. (2018) Localized Visualization and Autonomous Classification of Cell Surface Receptor Clusters Using DNA Proximity Circuit. *Anal. Chem.*, **90**, 6193–6198.
(2) Ang, Y.S., Bando, T., Sugiyama, H. and Yung, L.-Y.L. (2017) Dynamic Stabilization of Hybridization Chain Reaction Using Pyrrole-Imidazole Polyamide (Submitted).
(3) Ang, Y.S., Tong, R. and Yung, L.-Y.L. (2016) Engineering a Robust DNA Split Proximity Circuit With Minimized Circuit Leakage. *Nucleic Acids Res.*, **44**, e121.
(4) Ang, Y.S. and Yung, L.-Y.L. (2016) Rational Design of Hybridization Chain Reaction Monomers For Robust Signal Amplification. *Chem. Commun.*, **52**, 4219–4222.
(5) Ang, Y.S. and Yung, L.-Y.L. (2014) Engineering Self-Contained DNA Circuit For Proximity Recognition and Localized Signal Amplification of Target Biomolecules. *Nucleic Acids Res.*, **42**, 9523–9530.

(*Works not included in thesis*)

(6) Ang, Y.S., Woon, W.W.E. and Yung, L.-Y.L. (2018) Design of Spacer Sequence to Modulate Turn-On Fluorescence of DNA-Templated Silver Nanoclusters. *Nucleic Acids Res.*, gky521.
(7) Ang, Y.S. and Yung, L.-Y.L. (2016) In Kah, J. K. C. Y. and Xu, Z. J. (eds.), *New Developments in Gold Nanomaterials Research*. Nova Science Publishers, pp. 3–24.
(8) Ang, Y.S. and Yung, L.Y.L. (2014) Toehold-Mediated Internal Control to Probe the Near-Field Interaction Between Metallic Nanoparticle and Fluorophore. *Nanoscale*, **6**, 12515–12523.

A. Yan Shan, *Engineering a Robust DNA Circuit for the Direct Detection of Biomolecular Interactions*, Springer Theses,
https://doi.org/10.1007/978-981-13-2188-7

Patents Filed

Ang, Y.S. and Yung, L.-Y.L. NUS ILO Ref: 2016-219. A DNA-Based Probe System for the One-Pot, *in situ* Detection of Proximity Events Between Biomolecules.

Conferences

(1) Poster presentation on "Engineering a Robust DNA Split Proximity Circuit for the *in situ* Detection of Proximal Recognition Events" in the 22nd DNA Computing and Molecular Programming (DNA22 2016) in Munich, Germany.
(2) Poster presentation on "Strategies to Minimize DNA Circuit Leakage Across a Three-Way Junction" in the 22nd DNA Computing and Molecular Programming (DNA22 2016) in Munich, Germany.
(3) Oral presentation on "Design of Split DNA Assay for Evaluating Proximity-Based Biomolecular Interactions" in the 8th International Conference on Materials for Advanced Technologies (ICMAT 2015) in Singapore.
(4) Poster presentation on "Engineering Self-Contained DNA Circuit for Proximity Recognition and Localized Signal Amplification of Target Biomolecules" in the 20th DNA Computing and Molecular Programming (DNA20 2014) in Kyoto, Japan.
(5) Oral presentation on "Probing the true near-field interaction between metallic nanoparticle and fluorophore in bulk colloidal solution using toehold-mediated internal control" in UK Colloid 2014: An International Colloid and Interfacial Science Symposium in London, United Kingdom.
(6) Poster presentation on "High-yield synthesis of monodispersed gold nanostar with highly tunable optical property" in UK Colloid 2014: An International Colloid and Interfacial Science Symposium in London, United Kingdom.

MIX
Papier aus verantwortungsvollen Quellen
Paper from responsible sources
FSC® C105338

If you have any concerns about our products,
you can contact us on
ProductSafety@springernature.com

In case Publisher is established outside the EU,
the EU authorized representative is:
Springer Nature Customer Service Center GmbH
Europaplatz 3, 69115 Heidelberg, Germany

Printed by Libri Plureos GmbH
in Hamburg, Germany